Biomedical Statistics

Shakti Kumar Yadav · Sompal Singh ·
Ruchika Gupta

Biomedical Statistics

A Beginner's Guide

 Springer

Shakti Kumar Yadav
Pathology
NDMC Medical College and Hindu Rao
Hospital
New Delhi, India

Sompal Singh
Pathology
NDMC Medical College and Hindu Rao
Hospital
New Delhi, India

Ruchika Gupta
Cytopathology
ICMR-National Institute of Cancer
Prevention and Research
Noida, Uttar Pradesh, India

ISBN 978-981-32-9296-3   ISBN 978-981-32-9294-9   (eBook)
https://doi.org/10.1007/978-981-32-9294-9

This Springer imprint is published by the registered company Springer Nature Singapore Pte Ltd.
The registered company address is: 152 Beach Road, #21-01/04 Gateway East, Singapore 189721,
Singapore

Dedicated to our parents and teachers

Foreword

Biomedical statistics, though constituting a vital aspect of all research in the field, hardly gets its due attention in the medical education. This leads to many students and researchers feeling helpless during the compilation of their research with regard to the suitability as well as application of a statistical test to their data. The currently available books on biomedical statistics focus mainly on the application of the statistical tests without providing the knowledge on the fundamentals of the mathematical basis of these tests. In this regard, this book by the team led by Dr. Sompal Singh is a welcome addition to the literature on biomedical statistics.

Dr. Sompal Singh, a pathologist by profession, is an avid learner himself and a self-trained skilled statistician, who, for his love for the subject, pursued a graduate degree in statistics from the ICFAI University, Tripura, while continuing his job at the busy Hindu Rao Hospital, Delhi. His keen interest in mathematics is well reflected in the detailed explanations of the mathematical derivations of various formulae used in statistics.

Being a medical doctor and researcher himself, he has divided the book into four logical sections with a thoughtful flow of chapters and ample inclusion of figures, tables and exercises to help the readers understand the concept as well as the utility of each statistical test. One of the main merits of this book is the ease with which a student can apply the tests using a calculator without the need of elaborate statistical software. Keeping in mind the importance of conduct of epidemiological studies, sample size estimation and sampling methods have also been given their due place in the book.

This book by Yadav, Singh and Gupta is an easy-to-use "benchside" help in application of statistics by various biomedical students and researchers. I personally wish the team and their book all success.

Director Professor and Head   Jugal Kishore
Department of Community Medicine
VMMC & Safdarjung Hospital
(Ministry of Health and Family Welfare)
New Delhi, India

Editor in Chief (Hony),
Epidemiology International
and International Journal of Preventive,
Curative and Community Medicine
New Delhi, India

Preface

The subject of biomedical statistics is an integral part of the undergraduate training of medical and paramedical students and is used extensively by postgraduates in all fields of medicine in their theses and other research work.

Though many books on statistics are available today, those dealing with biomedical statistics are fewer in number. The available books on biomedical statistics usually cover only the application of the statistical tests. However, there is a lack of a book explaining the fundamentals of statistical tests or explaining the mathematical basis of derivation of statistical formulae.

This book on biomedical statistics explains the concept of statistical tests in such basic terms that a student can apply these tests using a calculator, a paper and a pen/pencil. Ample numbers of figures and tables have been included in the book to make it more explanatory and easier to grasp. In order to help the students, have clarity on the application of these tests, a number of solved exercises have been provided.

The book is divided into five logical sections, namely, Introduction with Descriptive Statistics, Tests for Inference (Hypothesis Testing), Demography and Life Tables and Basic Concepts of Probability. Since the readers of this book may wish to undertake an epidemiological study (case control or cohort) at some point in their career, the concepts of sampling methods and sample size calculation have also been explained. This is lacking in a few other books of biomedical statistics available in the market. Easy-to-use nonparametric tests are also included in a chapter since their application in biomedical statistics is ever increasing these days.

We hope that this book will be useful in the application of biomedical statistics in your research work and strengthen medical science through this endeavour.

New Delhi, India   Shakti Kumar Yadav
New Delhi, India   Sompal Singh
Noida, India   Ruchika Gupta

Acknowledgements

We gratefully acknowledge the guidance and encouragement provided by Prof. Jugal Kishore, the first teacher of statistics for Dr. Singh and a guiding force behind this book, without whose inputs, this book may have been incomplete. Our colleagues, Dr. Namrata Sarin, Dr. Sonam Kumar Pruthi and Dr. Aarzoo Jahan, and residents at Hindu Rao Hospital, Delhi, as well as Dr. Sanjay Gupta at the NICPR, Noida, deserve special mention for their constant support and help in every way in making the dream of this book a reality.

The strong support given by our families during the times we devoted ourselves completely to the writing and editing of this book helped us sail though such difficult times with ease. Last but not the least, thanks are due to our patients whose data we continuously analysed for checking and rechecking of the formulas in this book.

Contents

About the Authors

Shakti Kumar Yadav, MBBS, DCP and DNB Pathology, has been working as a Senior Resident at the Department of Pathology, North Delhi Municipal Corporation Medical College and Hindu Rao Hospital, New Delhi, since 2017. He completed his Diploma in Clinical Pathology at Gandhi Medical College, Bhopal, followed by DNB in Pathology at Hindu Rao Hospital, New Delhi. He has numerous publications to his credit and has been actively involved in teaching undergraduate and graduate medical and paramedical students.

Sompal Singh, BSc Statistics, MBBS, MD Pathology and MBA (Hospital Administration), has been working as a Senior Specialist at the Department of Pathology, NDMC Medical College and Hindu Rao Hospital since 2005. He completed his MD in Pathology at Maulana Azad Medical College, New Delhi, followed by senior residency at the same institute. He has 121 research publications in various national and international journals to his credit, and has also co-authored two books for paramedical students and one for medical undergraduate students. He has a keen interest in teaching, medical education research and breast pathology.

Ruchika Gupta, MBBS, MD (Pathology) and MBA (Hospital Administration), has been working as a Scientist at the Division of Cytopathology, ICMR-National Institute of Cancer Prevention and Research, Noida, since 2015. She completed her MD in Pathology at Maulana Azad Medical College, New Delhi, followed by a senior residency at AIIMS, New Delhi. She also trained in renal pathology during her stint as a pool officer at AIIMS. She also worked as an Assistant Professor and Head of the Department of Pathology, Chacha Nehru Bal Chikitsalaya, New Delhi, for nearly 5 years. She has 145 publications in various national and international journals to her credit, and has co-authored two books for paramedical students. She has been actively involved in organizing and conducting a series of workshops on cervical cancer screening for pathologists at the NICPR, which have attracted participants from across India and abroad.

Abbreviations

ANOVA	Analysis of Variance
CDF	Cumulative Density Function
CV	Coefficient of Variance
EPS	Equal Probability of Selection
ESM	Event Sampling Methodology
MoD	Measures of Dispersion
PDF	Probability Density Function
RCT	Randomized Control Trial
ROC	Receiver Operating Characteristics
SD	Standard Deviation
SE	Standard Error
SS	Sum of Squares
WHO	World Health Organization
WOR	Without Replacement
WR	With Replacement

Part I

Introduction to Statistics

Applications of Statistics

1

Abstract

Background
Applications of Statistics
    Use in anatomy and physiology
    Use in pharmacology
    Use in internal medicine
    Use in surgery
    Use in community medicine
    Use in lab medicine
    Use in teaching

Keywords

Biomedical statistics · Application of statistics

1.1 Background

Statistics is the science of figures. Biomedical statistics is a subdivision of this science dealing with statistics relevant to the medical science.

Grammatically, the word "statistics" used in a singular form represents a measured/counted piece of information where as its pleural form "statistics" represents a group of measured information. On the other hand, the word *Statistics* when used as a science means an act of gathering, compiling and analysing the data. The results of such data analysis utilizing statistics are used to draw conclusions and state recommendations about the population. Fig 1.1 illustrates the journey from data collection to recommendation.

Although Fig. 1.1 starts from data collection, work of statistician starts much before this step. At the onset of an activity involving statistical analysis, the statistician has to understand the hypothesis, calculate the sample size and devise the most

© Springer Nature Singapore Pte Ltd. 2019
S. K. Yadav et al., *Biomedical Statistics*,
https://doi.org/10.1007/978-981-32-9294-9_1

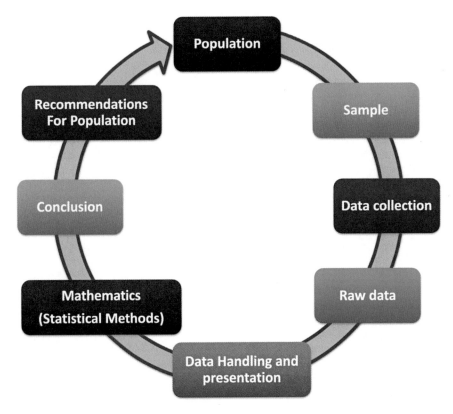

Fig. 1.1 Journey from data collection to recommendations

appropriate method of sample collection in order to be able to derive meaningful conclusions later.

1.2   Applications of Statistics

Statistical methods are required by all the divisions of medical science including anatomy, physiology, pharmacology, internal medicine, pathology and community medicine to name a few. The requirement varies according to the population under study and the objective of the mathematical procedures. A few common examples of the application of statistics in various specialities of medical science are mentioned below.

1.2.1   Use in Anatomy and Physiology

(a) To define the normal range of anatomical and physiological variables, e.g. designing growth charts which are used by paediatricians

(b) To assess the significance of physiological difference in certain parameters among individuals in different periods/places, e.g. difference in average haemoglobin between Asian and Western population

(c) To find the correlation between two variables in order to define normalcy, e.g. correlation between body height and weight which is used in interpretation of body mass index (BMI)

1.2.2 Use in Pharmacology

(a) During drug development, statistics is used in the assessment of the efficacy of a newly discovered drug as compared to placebo.

(b) Comparison of effect of two different drugs used for the same disease in order to frame clinical management guidelines.

1.2.3 Use in Internal Medicine

(a) Descriptive statistics is used to quantitate the signs and symptoms of a particular disease, e.g. a particular study revealing that "Disease A" presents with cough in 20%, fever in 10% and weight loss in 60% of patients.

(b) To find the association between exposure to a risk factor and occurrence of its related disease, e.g. smoking in lung cancer.

(c) To compare the clinical effect of two different treatment modalities available for the same disease.

(d) Application in survival studies, e.g. to identify 5-year survival rate after surgical management of a particular cancer.

(e) To identify etiological factors/risk factors/predictors of a given disease using multiple linear regression, path analysis, (multiple) logistic regression and discriminant function analysis.

1.2.4 Use in Surgery

(a) To compare the efficacy of two different surgical procedures for the same disease.

1.2.5 Use in Community Medicine

(a) In health planning, descriptive statistics (maternal mortality rate, infant mortality rate) of a community constitute an important data for health policymakers

(b) To assess the effectiveness of a public health measure, e.g. study of efficacy of chlorinated water in reducing the incidence of diarrhoeal disease as shown in Fig. 1.2

Fig. 1.2 Study of effectiveness of chlorinated water in society A and B

(c) To find the association between exposure to a risk factor and the likelihood of occurrence of a disease, e.g. to find association between radiation exposure to a body part and malignancy in the same organ (Fig. 1.3)

1.2.6 Use in Lab Medicine

(a) Establishment of reference ranges of haematological, biochemical, microbiological and serological assays

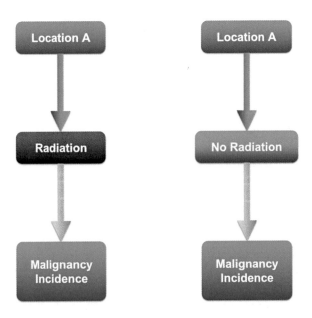

Fig. 1.3 Study to find association between radiation exposure and malignancy

(b) To draw individual laboratory reference ranges for certain parameters (e.g. reference range of Widal test results with respect to local population)

(c) Identification of prognostic markers: identify and quantify the correlation of cellular expression of various markers (proteins) with the diagnosis or prediction of prognosis of a disease, e.g. correlation of oestrogen and progesterone receptor expression in the tumour cells with prognosis of carcinoma breast

(d) To assess the sensitivity, specificity, negative predictive value and positive predictive value of a particular laboratory test

(e) Interobserver agreement for procedures involving an individual's observation and having clinical significance, e.g. interobserver agreement on grading of breast carcinoma

(f) Comparison of two methods giving numerical results (Bland and Altman graph)

(g) Analysis of time series and demand forecasting in laboratory management

(h) For assessment and interpretation of laboratory quality control methods – both internal and external quality assurance

1.2.7   Use in Teaching

(a) Utilization of learning curves for assessment of the teaching methods

The list of uses of statistics is not exhaustive by any means. In fact, statistics forms an integral and sometimes, an unrecognized part of medical practice.

Statistical Terms

2

Abstract

Statistical terms
Types of statistics
    Descriptive statistics
    Inferential statistics

Keywords

Descriptive statistics · Inferential statistics

2.1 Statistical Terms

Before learning about the individual statistical methods, we should apprise ourselves of the terms used in statistics and their exact meaning within the context of this branch of science (Table 2.1).

2.1.1 Variable

A variable is a *symbol* which is used for an entity that can assume any value or a set of values. It is a feature of interest in a study. Weight, height, age, haemoglobin and level of anxiety are some examples of a variable.

2.1.2 Value

It is the numerical value assigned to a particular variable. For example, if height is a variable, then Mr. XYZ's height of 173 cm is one of its values.

© Springer Nature Singapore Pte Ltd. 2019
S. K. Yadav et al., *Biomedical Statistics*,
https://doi.org/10.1007/978-981-32-9294-9_2

Table 2.1 Commonly
used statistical terms

Statistical terms
Variable
Value
Constant
Attribute
Population/universe
Sample
Statistics
Parameter

2.1.3   Constant

It is a symbol used to represent an entity having a fixed value irrespective of the context. For instance, π is a constant used in mathematics as well as statistics and has a value of 3.14.

2.1.4   Attribute

It is a *feature* which can take either nominal (that can be counted but not arranged in ascending or descending order) or ordinal (that having gradations) data type. Hence, usually attributes of a sample or population cannot be subjected to routine mathematical calculations like additions or taking average. For example, gender of study subjects is an attribute – we can count the number of individuals with a particular gender, but we cannot add the separate genders.

2.1.5   Population/Universe

Population is the entire group of subjects/individuals/items/objects for which a final statistical inference is drawn. It can be either finite or infinite.

2.1.6   Sample

A population is usually too vast to study as a whole. Hence, a sample, which is a small representative number of subjects/individual/items/objects of a particular population, is subjected to study, data collection and statistical analysis. Since a sample is meant to represent the entire population from which it has been derived, all efforts have to be undertaken to ensure that the sample is of adequate size, unbiased and truly representative of the population, so that the inferences drawn from this sample can be extrapolated to the population it has been derived from.

2.1.7   Statistics

It is the measurable quantity which is obtained from a sample.

2.1.8   Parameter

Parameter is the feature of a population that can be estimated with reasonable uncertainty from sample statistics (Table 2.2). Since the statistics is derived from a sample, it usually is not equal to the population parameter, and this difference is known as *sampling error.*

These parameter symbols shall be used in the various formulas throughout the book.

2.2   Types of Statistics

Statistics can be divided into two types: descriptive statistics and inferential statistics.

2.2.1   Descriptive Statistics

Descriptive statistics describes the features of a particular sample. In this, no comparison is made between groups nor the significance of difference is assessed, although confidence level is stated with some descriptive statistics. For example, in a particular study with two groups, the descriptive findings are shown in Tables 2.3 and 2.4.

Table 2.2 Difference between sample statistics and population parameter

Name	Statistics (of sample)	Parameter (of the population)
Mean	$\bar{x}$	μ
Standard deviation	s	σ
Variance	s^2	σ^2
Proportion	p	P

Table 2.3 Descriptive statistics of group 1

Group 1 Hb%	Descriptive statistics
10	Number of samples: 7
12	Range: 8–14
13	Median (middle value): 11
10	Mean: 11.14
8	
14	
11	

Table 2.4 Descriptive
statistics of group 2

Group 2 Hb%	Descriptive statistics
11	Number of samples: 7
12	Range: 11–25
13	Median (middle value): 13
11	Mean: 14.14
13	
14	
25	

Descriptive statistics have a number of applications in the medical field like:

1. Identification of a pattern in a set of data.
2. To identify outliers in a set of values.
3. Guidance of choice of further statistical tests to be applied on the set of data.
4. It leads to the generation of a hypothesis. For example, by simply looking at Table 2.3 and 2.4, it can be hypothesized that mean of group 2 is more than group 1.

2.2.2   Inferential Statistics

Inferential statistics assesses the significance of the result of statistical tests applied. For instance, the significance of difference in a variable between two samples from a population or between two populations can be derived from inferential statistics. It helps to distinguish true difference from random variation in a sample or population. Inferential statistics hence allows testing of the hypothesis generated by descriptive statistics.

For example, the data shown in Tables 2.3 and 2.4 generates a hypothesis that the mean of group 2 is higher than group 1. However, application of an appropriate statistical test like Student's t-test (described in later chapters) reveals that the difference in mean of the two groups was statistically insignificant (p-value $= 0.16$). Hence, the hypothesis is proven wrong using inferential statistics.

Data Types

3

Abstract

Background
Categorical or Qualitative data
    Nominal data
    Ordinal data
Numerical data
    Discrete data
    Continuous data
Cross-sectional data
Time Series data

Keywords

Nominal data · Ordinal data · Discrete data · Continuous data

3.1 Background

Data is any information obtained from the sample of a population or from census. It may be quantitative or qualitative in nature. Knowledge of the types of data (Table 3.1) is paramount in understanding of the appropriate statistical test to be applied.

3.2 Categorical or Qualitative Data

Categorical data is one that by itself cannot take a numerical value and, hence, cannot be subjected to mathematical calculations. There are two types of categorical data: nominal and ordinal described below.

© Springer Nature Singapore Pte Ltd. 2019
S. K. Yadav et al., *Biomedical Statistics*,
https://doi.org/10.1007/978-981-32-9294-9_3

Table 3.1 Types of data

Data types
Categorical or qualitative data
Nominal data
Ordinal data
Numerical data
Discrete data
Continuous data
Cross-sectional data
Time series data

3.2.1 Nominal Data

Nominal data is a type of qualitative data which describes the quality or characteristics (attribute) of a variable. It is usually composed of labels or names that can be counted. However, the labels cannot be arranged in ascending or descending order.

For example:

- Gender of osteosarcoma patients (male, female)
- Smoking habits of lung cancer patients (yes, no)
- Tobacco chewing habits of patients with oral leucoplakia (yes, no)

3.2.2 Ordinal Data

This type of qualitative data has gradations and hence can be arranged in a meaningful order, i.e. either ascending or descending order. Ordinal data is usually coded like "1,2,3…so on" and "I, II, III…so on".

For example:

- Staging of carcinoma (stage I, II, III, IV)
- Grading of inflammation (mild, moderate, severe)
- Grading of dysplasia (mild, moderate, severe)

3.3 Numerical Data

In contrast to categorical data, the numerical data takes a quantitative value. Such a value can be counted and arranged in an order. In addition, numerical data can also be subjected to various mathematical calculations. Numerical data can be of two types: discrete and continuous.

3.3.1 Discrete Data

It is a type of quantitative data which can take only integer values (no fractions) and is not continuous. However, it can be taken up as "continuous data" if there are sufficient number of possible values, else it should be taken up as ordinal data. For example if discrete data ranges from one to four, then it is better to take it as ordinal data for further statistical handling. Whereas, if discrete data ranges from 1 to 1000, then it can be taken as continuous data and can be subjected to all statistical tests suitable for continuous data type. Discrete data is described by its median and the interquartile range.

For example:

- Number of children in various standards or classes in a school
- Number of patients with fever attending OPD daily over a period of 30 days

3.3.2 Continuous Data

This type of quantitative data is continuous and, hence, can take any fractional value. It is described by mean and standard deviation.

For example:

- Age of leukaemia patients
- Haemoglobin levels of school children
- Serum bilirubin levels of patients with acute viral hepatitis

A summary of the data types described so far is presented in Table 3.2.

There is another way of classification of data not related to the aforementioned classification. Data can be categorized as cross-sectional data and time series data.

Table 3.2 Summary of data types

Categorical data (qualitative)		Numerical data (quantitative)	
Nominal data	Ordinal data	Discrete data	Continuous data
Countable but cannot be arranged in order	Has gradations and can be arranged in order	Data is not continuous and can take integer values	Data is continuous and can take any fractional value
For example, gender of students of a particular class	For example, stages of breast cancer	For example, number of patients attending OPD on a particular day	For example, haemoglobin values of school children

3.4   Cross-Sectional Data

Data collected at a point of time or over a small time period is termed as cross-sectional data. This type of data is frequently collected in epidemiological studies.

3.5   Time Series Data

Data collected at different points of time or different periods of time for a single variable is classified as time series. This type of data is very useful in analysing time trends for the studied variable in a sample or population.

Data Classification

4

Abstract

Background
Classification of numerical data
Classification of categorical data
    Frequency table
    Contingency table/Cross-tabulation

Keywords

Class interval · Frequency table · Contingency table · Cross-tabulation

4.1 Background

Raw data that is collected from a sample or population cannot usually be subjected to statistical analysis directly. First, the data needs to be classified for visual under-standing and presentation of data for further statistical calculations. The method of data classification varies with the type of data.

4.2 Classification of Numerical Data

Numerical data can be easily classified in the form of *frequency tables*, which are simply tabulations of the number samples in each defined category of a variable. There is a slight difference in the frequency tables used for discrete or continuous data. Discrete data, that we learnt in Chap. 3, can be classified by noting the frequencies for each and every value of data as illustrated in Table 4.1. On the other hand, values of continuous data are better depicted when they are classified in the form of class intervals (Table 4.2). This is so because continuous data is allowed to have fractional values as well. In such a situation, it becomes cumbersome to make a table depicting frequencies for each value of data.

© Springer Nature Singapore Pte Ltd. 2019
S. K. Yadav et al., *Biomedical Statistics*,
https://doi.org/10.1007/978-981-32-9294-9_4

Table 4.1 Frequency table of discrete data – number of children in a study of 168 families

Number of children	Number of families
0	3
1	50
2	75
3	25
4	10
5	5

Table 4.2 Example of classification of continuous data – haemoglobin values in a study of 145 patients

Haemoglobin (gm/dl)	Number of patients
5–7	5
7–9	25
9–11	35
11–13	80
13–15	31

There Are Certain General Rules for Making Class Intervals

1. The number of classes should ideally be between 5 and 15 depending on the size of the sample. If the number of classes becomes more than 15, the class interval should be increased to reduce the number of classes.
2. The class interval should neither be too wide nor too narrow.
3. The class intervals preferably should be same for all the classes, so that the data in these classes can be compared.
4. Various classes should be arranged in a meaningful order.
5. The upper limit of classes may be inclusive or exclusive. For instance, in Table 4.2, the upper limit of classes is inclusive since there is overlap between upper limit of one class and lower limit of the consecutive next class. In Table 4.2, first class actually range from 5 to 6.999; similarly, second class range from 7 to 8.999 and so on.

Exercise 4.1

Height (in cm) of 40 students studying in first year of medical school is given as follows:

135, 145, 147, 176, 145, 135, 172, 169, 165, 149, 144, 160, 167, 139, 141, 160, 156, 159, 153, 158, 146, 168, 161, 151, 143, 175, 147, 146, 179, 162, 138, 164, 155, 151, 165, 144, 147, 148, 141, 178.

Classify the data with appropriate class intervals.

Solution (Table 4.3)

$$\text{Range} = \text{highest value} - \text{lowest value} = 179 - 135 = 44$$

Table 4.3 Frequency table
for the data given in
exercise 4.1

Class interval	Frequency
131–140	4
141–150	14
151–160	9
161–170	8
171–180	5
Total	40

To make five classes, we require a class interval of 44/5 = 8.8.

Hence, we can conveniently make five classes starting from 131 with a class interval of 10.

4.3   Classification of Categorical Data

Categorical data can be classified in the form of frequency table or contingency table/cross-tabulation.

4.3.1   Frequency Tables

The frequency table used for categorical data differs from the one for numerical data in that here, the number of individual/items possessing a particular attribute is written in front of that attribute. Since the data is in the form of attributes which are usually mutually exclusive, proportions can also be calculated as shown in Table 4.4.

4.3.2   Contingency Table/Cross-Tabulation

In cross-tabulation, the number of individuals/items is tabulated across two different attributes as shown in Table 4.5. This is the most appropriate data classification method when two different but possibly related attributes of a sample need to be presented and analysed simultaneously.

Exercise 4.2

In a particular study, liver disease was found in 350 out of 500 subjects in the study group with history of chronic alcohol intake, compared to 50 out of 1000 subjects in control group without history of alcohol intake. Classify this data using cross-tabulation with respect to alcohol intake and liver disease.

Table 4.4 Tabulation of
gender details of a study

Category	Frequency	Proportion
Males	25	25/61
Females	36	36/61
Total	61	61/61

Table 4.5 Cross-tabulation of data for association of smoking and lung cancer

	Smoking – Yes	Smoking – No	Total
Lung cancer – Yes	25	75	100
Lung cancer – No	5	95	100
Total	30	170	200

Solution (Table 4.6)

Table 4.6 Cross-tabulation for data given in exercise 4.2

	Alcohol intake – Yes	Alcohol intake – No	Total
Liver disease – Yes	350	50	400
Liver disease – No	150	950	1100
Total	500	1000	1500

Data Presentation

5

Abstract

Background
Pictogram
Pie diagram
Bar diagram
Line diagram
Histogram
Frequency polygon
O'give curve
Scatter diagram
Regression line
Box and Whisker plot

Keywords

Bar diagram · Histogram · Scatter diagram · Regression line · Box and Whisker plot

5.1 Background

Data presentation is an important aspect of mathematical science to provide appropriate visual impression of the data. This can be done in the form of pictures, diagrams or graphs. Given the multiplicity of choices of data presentation methods with somewhat overlapping usages, selection of the most appropriate graph type assumes importance for effective presentation of a given data.

© Springer Nature Singapore Pte Ltd. 2019
S. K. Yadav et al., *Biomedical Statistics*,
https://doi.org/10.1007/978-981-32-9294-9_5

Fig. 5.1 Pictogram representing an epidemiological study showing number of men, women and children having normal haemoglobin (each picture denotes ten individuals in that category)

5.2   Pictogram

A *pictogram* depicts data in the form of pictures using different graphics for various categories. The frequencies of the individual categories are represented by appropriate number of pictures where each picture denotes a certain predefined quantity (5, 10, 15 etc.), as shown in Fig. 5.1.

Although such a data can also be presented by using bar diagram (see below), pictograms are very useful to explain your data to those readers who are not well-versed with statistical graphs. This type of tool is especially useful for representation of data in public media such as newspapers and magazines designed for the general population to enable easy comprehension.

In medical practice, pictograms can be used for explaining relevant information to the general public or patients who are not very well educated or during field surveys or health education camps.

5.3   Pie Diagram

In a *pie diagram*, the data is represented in the form of sectors of a circle, where the angle of a particular sector is proportional to its frequency. It is used to depict composition or proportions and helps in getting an immediate visual impression of the predominant proportion as well as the lowest contributing proportion. The data for pie chart could be absolute numbers or in percentages (Fig. 5.2).

Exercise 5.1
In an individual, the composition of serum proteins is as follows: albumin 2.91 g/dl, α-globulin 1.13 g/dl, β-globulin 1.55 g/dl, γ-globulin 1.02 g/dl. Make a suitable diagram for representation of this data.

Solution
Let us first tabulate the given information and calculate the percentage composition of each component (Table 5.1).

Fig. 5.2 Distribution of
students according
to academic grades in high
school (class 10th)

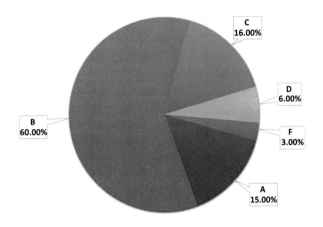

Table 5.1 Composition of
serum proteins.

Protein	Values (g/dl)	Percentages
Albumin	2.91	44.02
α-globulin	1.13	17.10
β-globulin	1.55	23.45
γ-globulin	1.02	15.43
Total	6.61	100

Next, a pie chart can be prepared using either the absolute values or the
percentages calculated thereof (Fig. 5.3).

5.4   Bar Diagram

This is the most frequently used data representation tool. In this method, the data is
represented in the form of bars, where height of each bar corresponds to the frequency
of the corresponding category. This is the conventional bar diagram (Fig. 5.4).

Various modifications have also been incorporated in the conventional bar
diagram, detailed below:

- *Bar diagrams with stalks* are similar to the conventional bar diagram with
  addition of stalks. Apart from the frequency depicted by the bars, this type of
  graph also depicts the range, variance or standard deviation of the included data in
  the form of stalks (Fig. 5.5).

Fig. 5.3 Pie chart for the data
given in Table 5.1.

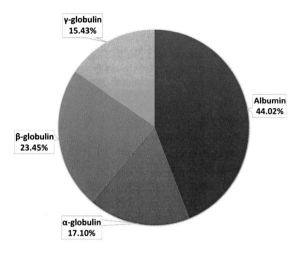

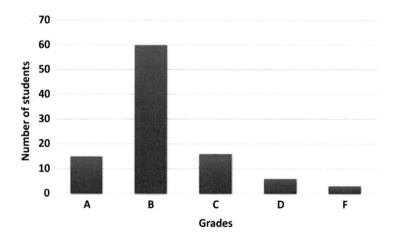

Fig. 5.4 Academic grades of students studying in high school (class 10th)

- *Composite bar diagrams* are used for data that can be classified into separate categories, each with similar subcategories. The proportion of these subcategories can be depicted and visually compared using the composite bar diagram (Fig. 5.6).

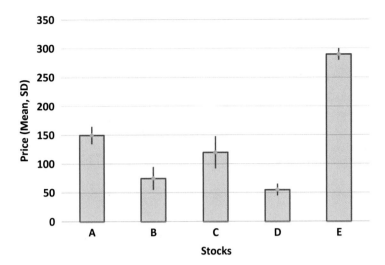

Fig. 5.5 Bar diagram with stalks showing mean stock market prices and standard deviation of five companies/stocks

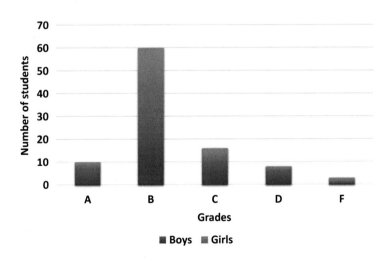

Fig. 5.6 Composite bar diagram showing gender distribution of each academic grade in high school students

Exercise 5.2

In a particular study, the incidence rate of HIV infection in various states of India shown in Table 5.2:

Draw a suitable diagram to show comparison of incident HIV cases between the states.

Solution
The given data can easily be depicted using a conventional bar diagram as below (Fig. 5.7).

Table 5.2 Incidence of HIV infection

State	New cases identified per 10,000 uninfected individuals
Delhi	3
Goa	4.2
Manipur	14.3
Mizoram	20.4
Maharashtra	3.3
Andhra Pradesh	6.3
Telangana	7
Karnataka	4.7
Nagaland	11.5

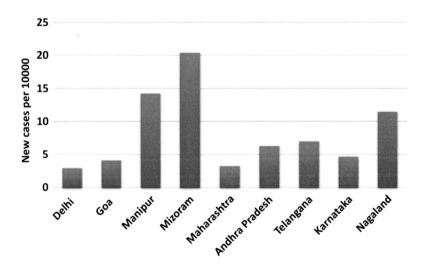

Fig. 5.7 Incidence rate of HIV infection of data given in Table 5.2

Table 5.3 Gender distribution of various age groups

Age groups (in years)	Male	Female
20–30	1	9
30–40	2	25
40–50	4	55
50–60	4	60
60–70	1	42
70–80	1	21

Exercise 5.3
In a particular study on breast cancer, the age and gender distribution of the patients is shown in Table 5.3.

Draw a suitable diagram.

Solution
Since this data has categories and subcategories, the best way of presentation would be a composite bar diagram (Fig. 5.8).

Exercise 5.4
In a clinical trial, continuous diastolic blood pressure monitoring was done after administration of five experimental drugs in five separate study groups of hypertensive patients. The results of the study are given in Table 5.4.

Draw a suitable diagram.

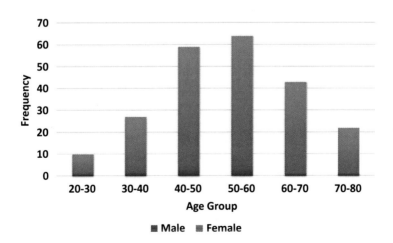

Fig. 5.8 Distribution of age groups and gender in breast cancer patients of data in Table 5.5

Table 5.4 Descriptive statistics of various drugs

Drug	Mean − 2SD	Mean	Mean + 2SD
A	76	84	92
B	84	96	108
C	96	100	104
D	76	82	88
E	80	88	96

Solution

In this example, bar diagram with stalks would be suitable since the standard deviation can also be depicted (Fig. 5.9).

5.5   Line Diagram

Line diagram is quite similar to the bar graph, but instead of drawing bars, a line that connects the points corresponding to the height of bars is drawn. Line diagram is generally used to represent trends over a period of time (Fig. 5.10) or gradations of a data. This tool gives an advantage of allowing trend comparison between different categories of data in the same graph.

Unlike bar diagrams, the categories on x-axis in a line diagram are arranged in a meaningful order, either in an ascending or descending order, for an easy visual impression of the trend across the categories.

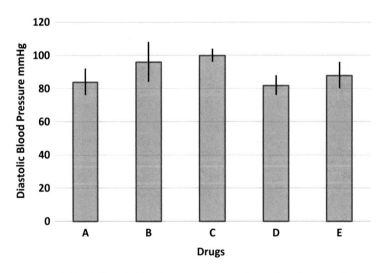

Fig. 5.9 Mean and SD of diastolic blood pressure after administration of various drugs

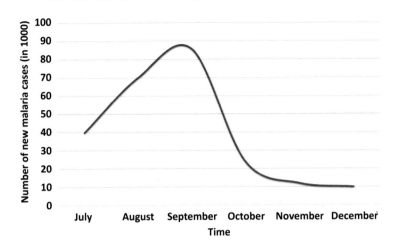

Fig. 5.10   Incidence of malaria in Delhi from July to December

Table 5.5   Mitotic count in various WHO grades

WHO grade	Mitotic count (per HPF)
I	2
II	7
III	20
IV	35

Exercise 5.5

The mitotic count observed in a particular study among various WHO grades of central nervous system astrocytic tumours are shown in Table 5.5.

Draw a suitable diagram illustrating the trend of mitotic count across WHO grades

Solution

The trend of mitotic count across WHO grades of astrocytic tumours is depicted in a line diagram (Fig. 5.11).

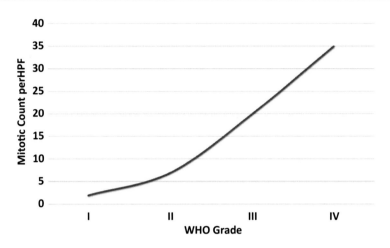

Fig. 5.11 Observed mitotic counts

5.6   Histogram

Histogram looks similar to bar diagram because of the use of bar-like figures. However, a histogram is entirely different from a bar graph because the x-axis of the histogram denotes class interval instead of discrete categories. Since the class intervals are continuous, the bars in a histogram are opposed to or touching each other. The width of each bar in a histogram represents a particular class interval, and its height corresponds to the frequency of that class interval. Hence histogram is the plot of frequencies (y-axis) against the class interval (x-axis). For instance, the frequency (y-axis) of haemoglobin values (as class intervals) of 145 patients in a study is shown in Fig. 5.12.

Exercise 5.6
In a study on breast cancer, the number of cases in various age groups is given in Table 5.6.

Draw a suitable diagram.

Solution
Given that the data is shown in class intervals, histogram can be drawn to depict this data (Fig. 5.13).

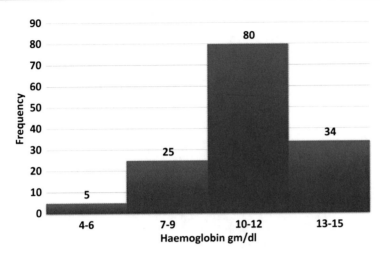

Fig. 5.12 Haemoglobin values in a study of 145 patients

Table 5.6 Age distribution of breast cancer patient	Age groups (in years)	Frequency
	20–30	4
	30–40	15
	40–50	10
	50–60	12
	60–70	8
	70–80	2

5.7   Frequency Polygon

A *frequency polygon* is obtained by drawing a line joining frequencies of classes or joining the highest points of the bars in corresponding histogram (Fig. 5.14). This is similar to the relationship between line diagram and bar graph.

5.8   O'give Curve

Here cumulative frequencies are plotted across the various class intervals. If accumulation starts from the beginning, it is known as *less than O'give* (Fig. 5.15), whereas, if accumulation starts from the end in reverse manner, it is known as *more than O'give* (Fig. 5.16).

To understand the concept and application of O'give curve, consider the data in Table 5.7.

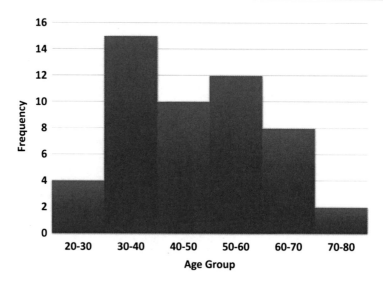

Fig. 5.13 Histogram of age groups of breast cancer patients

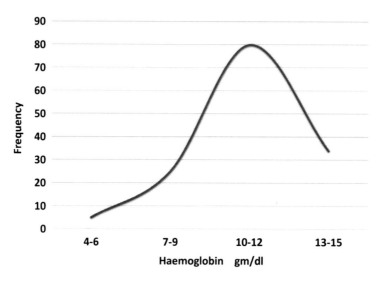

Fig. 5.14 Frequency polygon of data shown in Fig. 5.12

The third column in Table 5.7 is the cumulative usage of reagents which on plotting results in a less than O'give curve (Fig. 5.17). The rapid rise of the graph denotes that about 20% of items are responsible for about 80% total usage (Pareto principle, as used in inventory management). The main use of O'give curve is in the clinical laboratory for effective inventory management.

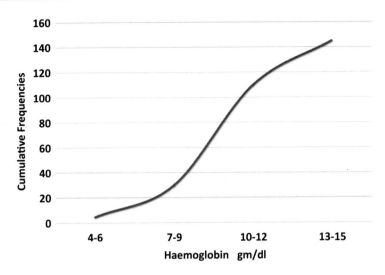

Fig. 5.15 Less than O'give curve for data shown in Fig. 5.12

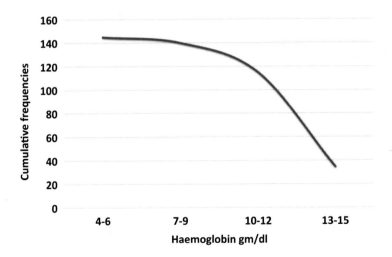

Fig. 5.16 More than O'give curve for data show in Fig. 5.12

Table 5.7 Annual reagent consumption in a clinical laboratory

Reagent name	Annual usage (USD)	Cumulative usage (USD)	Cumulative percentage (%)
Glucose	40,000	40,000	73.39
TSH	8000	48,000	88.07
Widal	4000	52,000	95.41
Uric Acid	1500	53,500	98.17
Calcium	800	54,300	99.63
Haemoglobin	200	54,500	100.00

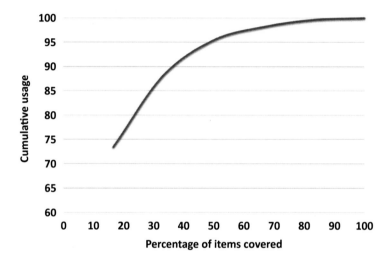

Fig. 5.17 Plot of cumulative usage of reagents

5.9   Scatter Diagram

This is frequently used to depict data representing correlation between two variables. In a scatter diagram, paired values of two variables are plotted on *x*- and *y*-axis. Each dot represents a paired value. In situations where there is no correlation between the variables, the dots are randomly scattered in the plot area. When a correlation is seen between two variables, the dots approximately lie in a linear fashion with either an upward pattern (positive correlation, Fig. 5.18) or a downward pattern (negative correlation).

5.10   Regression Line

Regression is used to determine the relationship between two variables – a dependent and an independent variable. Once the regression line is plotted, the value of dependent variable can be estimated from the value of independent variable though a derived equation. When there are only two variables in consideration, it is known as

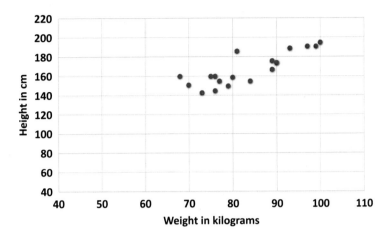

Fig. 5.18 Scatter diagram showing correlation between height and weight of students in college

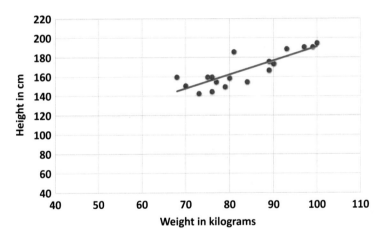

Fig. 5.19 Regression of weight overheight

univariate regression, since one is dependent and the other is independent variable. If there are more than one independent variables, then it becomes a multiple regression.

When one variable is regressed over the other, an equation of graph is obtained. That graph may be linear (straight line) or non-linear (polygon). Regression is meaningful when there is a correlation between the independent and dependent variables. Hence, in Fig. 5.19 which is the extension of Fig. 5.18 of correlation, we overlaid a regression line to explain this concept.

Exercise 5.6

The haemoglobin (gm/dL) and RBC count (million/µL) values of 19 subjects in a study are given in Table 5.8.

Draw a suitable diagram to illustrate the correlation (if any) between haemoglobin and RBC count and also draw a regression line.

Solution

First, a scatter plot of haemoglobin (x-axis) and RBC count (y-axis) is to be plotted. Then, a regression line can be overlaid (Fig. 5.20).

Table 5.8 Hemoglobin values with their corresponding RBC counts of 19 subject

Hemoglobin	RBC count
10.2	3.6
10	3.3
12.3	4.3
12.6	4.2
15.7	5.7
11.5	3.9
11	3.9
11.5	3.5
12.1	4.8
12.6	4.9
12.5	4.2
12.7	4.5
12.3	3.9
11.8	3.7
11.9	3.6
10.7	3.4
10.2	3.1
13.1	4.2
13.2	4.5

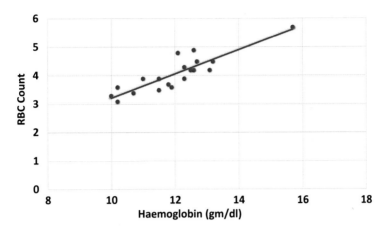

Fig. 5.20   Scatter and regression line of the data given in exercise 5.6

5.11   Box and Whisker Plot

It is used to represent a data which is not normally distributed (explained in Chap. 10), i.e. the curve of distribution of data is not bell-shaped. Here 2.5th, 25th, 50th (median), 75th and 97.5th percentile is plotted (Fig. 5.21; see Chap. 7 for calculation of percentiles).

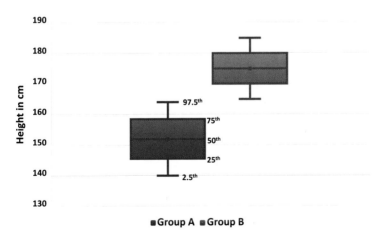

Fig. 5.21   Comparison of height of group A and B

Exercise 5.7

The marks obtained by students of section A and B in a particular subject are shown in Table 5.9.

Table 5.9 Marks of students in section A and B

Section A	Section B
30	11
41	12
58	15
45	16
48	14
52	17
52	18
78	13
52	12
53	11
63	14
54	15
51	16
60	5
52	4
53	5
56	2
57	3
54	5
58	8
65	5
56	7
58	5
57	4
52	5
51	7
54	5
52	3
52	8
51	30

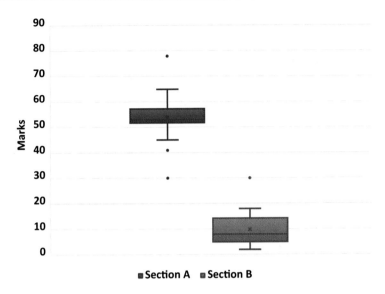

Fig. 5.22   Box and Whisker plot of data given in exercise 5.7

Draw a suitable diagram for visual comparison of the marks along with outliers.

Solution (Fig. 5.22)

Measures of Central Tendency

6

Abstract

Background
Mean
Median
Mode

Keywords

Central tendency · Mean · Median · Mode

6.1 Background

Data can be described by its spread as well as the tendency to cluster around a central value. This tendency of the data to cluster around a central value is known as "central tendency". There are three statistics for the measure of central tendency: *mean, median and mode* (Table 6.1).

6.2 Mean

It is also commonly known as arithmetic mean when no classifier is used (other means include geometric mean and harmonic mean which are rarely used in biomedical statistics). Arithmetic mean is obtained by summation of all the observations and dividing the resulting total by the number of observations (sample size). Various formulas for calculation of mean are as follows:

© Springer Nature Singapore Pte Ltd. 2019
S. K. Yadav et al., *Biomedical Statistics*,
https://doi.org/10.1007/978-981-32-9294-9_6

Table 6.1 Measures of central tendency

Measures of central tendency
Mean (arithmetic, geometric, harmonic)
Median
Mode

1. For raw data

$$\text{Mean} = \frac{\Sigma x}{n} \qquad \text{(Formula 6.1)}$$

2. For classified data

$$\text{Mean} = \frac{\Sigma f_i x_i}{\Sigma f_i} \qquad \text{(Formula 6.2)}$$

Where x_i is either the value of the ith class or the mid-point of the ith class interval and f_i is the frequency of ith class.

3. For classified data with available probabilities

$$\text{Mean} = \sum p(x_i) * x_i \qquad \text{(Formula 6.3)}$$

where $p(x_i)$ is probability of ith value (x_i) Formula 6.3 will be explained later with mathematical expectation in Chap. 26.

Let us now understand the calculation of mean through some exercises.

Exercise 6.1

There are 30 students in a class. The marks obtained by students in Anatomy test are given as: 10, 15, 58, 45, 48, 52, 52, 78, 52, 53, 89, 54, 51, 85, 52, 53, 56, 57, 54, 58, 98, 56, 58, 57, 52, 51, 54, 52, 52, 51. Calculate the mean of the given data set and draw a suitable diagram.

Solution

$$n = 30$$

Mean using Formula 6.1.

$$\bar{x} = \frac{\Sigma x}{n}$$

$$\bar{x} = \frac{\begin{array}{c} 10 + 15 + 58 + 45 + 48 + 52 + 52 + 78 + 52 + 53 + 89 + 54 + 51 + 85 + \\ 52 + 53 + 56 + 57 + 54 + 58 + 98 + 56 + 58 + 57 + 52 + 51 + 54 + 52 + 52 + 51 \end{array}}{30}$$

$$\bar{x} = 55.1$$

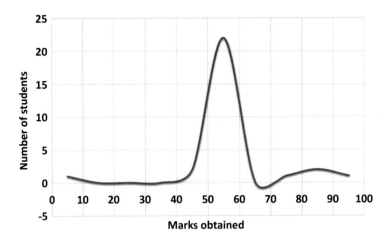

Fig. 6.1 Graphical representation of classified data shown in exercise 6.1

As can be seen in Fig. 6.1, the data is normally distributed with a bell-shaped curve. Hence, the calculated mean lies at the mid-point of the graph.

Exercise 6.2
The marks obtained by 30 students in a pharmacology test are given as:

$$1, 2, 5, 6, 4, 7, 8, 3, 2, 1, 4, 5, 6, 5, 4, 5, 2, 3, 5, 8, 5, 7, 5, 4, 5, 7, 5, 3, 8, 98$$

Calculate the mean of given data set and depict in a diagram.

Solution

$$n = 30$$

Using Formula 6.1

$$\bar{x} = \frac{\Sigma x}{n}$$

$$\bar{x} = \frac{\begin{array}{c}1 + 2 + 5 + 6 + 4 + 7 + 8 + 3 + 2 + 1 + 4 + 5 \\ + 6 + 5 + 4 + 5 + 2 + 3 + 5 + 8 + 5 + 7 + 5 + 4 + 5 + 7 + 5 + 3 + 8 + 98\end{array}}{30}$$

$$\bar{x} = \frac{233}{30}$$

$$\bar{x} = 7.76$$

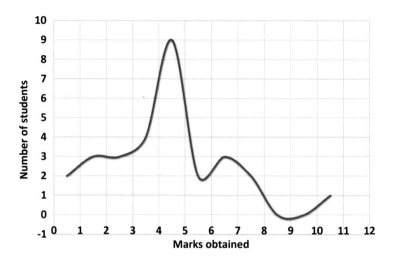

Fig. 6.2 Graphical representation of classified data shown in exercise 6.2

As illustrated in Fig. 6.2, although the mean is considered as a measure of central tendency, red arrow representing mean in this case is nowhere near the centre of data. This is due to the effect of outliers (values in the extremes of range) in the data shown in exercise 6.2. The data in this exercise is not normally distributed, and hence, the mean does not provide a true picture of the central tendency of data.

Exercise 6.3
In a class of 120 students, the height of students was measured, and data was compiled as given in Table 6.2. Calculate the mean height of the students.

Solution
Using Formula 6.2. Calculations are depicted in Table 6.3.

Table 6.2 Height distribution of 120 students

Height in cms	Number of students
100–110	11
111–120	19
121–130	31
131–140	21
141–150	18
151–160	12
161–170	8

Table 6.3 Frequency table for calculation of mean of classified data in exercise 6.3

Height in cms	Frequency (f_i)	Mid-point of the ith class interval (x_i)	$f_i x_i$
100–110	11	105	1155
111–120	19	115	2185
121–130	31	125	3875
131–140	21	135	2835
141–150	18	145	2610
151–160	12	155	1860
161–170	8	165	1320
Total	$\Sigma f_i = 120$		$\Sigma f_i x_i = 15{,}840$

$$\text{Mean} = \frac{\Sigma f_i x_i}{\Sigma f_i}$$

$$\text{Mean} = \frac{15840}{120}$$

$$\text{Mean} = 132$$

6.2.1   Salient Features of Mean

1. Since it is mathematically calculated, mean can be incorporated in further complex mathematical formulae and calculations.
2. Its calculation includes each and every observation of the data.
3. Since mean includes each observation, it is affected by extreme values, as we have seen in exercise 6.2.
4. It is not recommended for description of the data, which is not *normally distributed* (see Chap. 10).

6.3   Median

Median is the middle observation obtained after arranging the data in either ascending or descending order. The formulas for estimation of median are:

1. For even sample size

$$\text{Median} = \text{Mean} \left\{ \left[\frac{n}{2}\right]^{th} \text{value} + \left[\frac{n+2}{2}\right]^{th} \text{value} \right\} \qquad \text{(Formula 6.4)}$$

2. For odd sample size

$$\text{Median} = \left[\frac{n+1}{2}\right]^{th} \text{value} \qquad\qquad \text{(Formula 6.5)}$$

3. For classified data

$$\text{Median} = LL_m + \left[\frac{\Sigma f_i}{2} - CF_{m-1}\right] * \frac{CI_m}{f_m} \qquad\qquad \text{(Formula 6.6)}$$

where:

m is the median class or the class in which the median cumulative frequency lies $(\frac{\Sigma f_i}{2})$.
LL_m is the lower limit of the median class.
f_i is the frequency of the ith class.
CF_{m-1} is the cumulative frequency of the class preceding the median class.
f_m is the frequency of the median class.
CI_m is the class interval of the median class.
The estimation of median shall be better understood with the following examples.

Exercise 6.4
For the data given in exercise 6.1, estimate the median of marks obtained by the 30 students.

Solution
To identify the median, the data is first arranged in ascending order as follows:

10, 15, 45, 48, 51, 51, 51, 52, 52, 52, 52, 52, 52, 52, <u>53, 53</u>, 54, 54, 54, 56, 56, 57, 57, 58, 58, 58, 78, 85, 89, 98

$$n = 30$$

$$\text{Median} = \text{Mean} \left\{ \left[\frac{n}{2}\right]^{th} \text{value} + \left[\frac{n+2}{2}\right]^{th} \text{value} \right\}$$

$$\text{Median} = \text{Mean} \left\{ 15^{th} \text{value} + 16^{th} \text{value} \right\}$$

$$\text{Median} = \text{Mean} \left\{ 53 + 53 \right\}$$

$$\text{Median} = 53$$

Note that in this example as the data is normally distributed, mean and median are close to each other (Fig. 6.3). Hence, either of them can be used as a measure of central tendency.

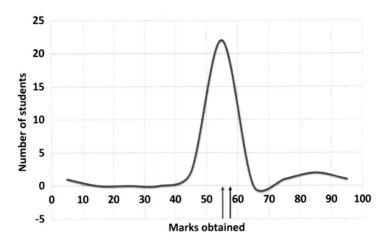

Fig. 6.3 Median (blue arrow) and mean (red arrow) of data shown in exercise 6.1

Exercise 6.5

For the data given in exercise 6.2, estimate the median marks.

Solution

To estimate the median, the data is arranged in ascending order as follows:

$$1, 1, 2, 2, 2, 3, 3, 3, 4, 4, 4, 4, 5, 5, \underline{5, 5}, 5, 5, 5, 5, 5, 6, 6, 7, 7, 7, 8, 8, 8, 98$$

$$n = 30$$

$$\text{Median} = \text{Mean}\left\{ \left[\frac{n}{2}\right]^{\text{th}} \text{value} + \left[\frac{n+2}{2}\right]^{\text{th}} \text{value}\right\}$$

$$\text{Median} = \text{Mean}\left\{15^{\text{th}} \text{ value} + 16^{\text{th}} \text{ value}\right\}$$

$$\text{Median} = \text{Mean}\left\{5 + 5\right\}$$

$$\text{Median} = 5$$

Exercise 6.6

For the data given in exercise 6.3, estimate the median height.

Solution

To estimate the median height using the Formula 6.6. Initial calculations are shown in Table 6.4.

Table 6.4 Frequency distribution of height

Height in cms	Frequency (f_i)	Cumulative frequency
100–110	11	11
111–120	19	30
121–130	31	61
131–140	21	82
141–150	18	100
151–160	12	112
161–170	8	120
Total	$\Sigma f_i = 120$	

$$\text{Median} = LL_m + \left[\frac{\Sigma f_i}{2} - CF_{m-1}\right] * \frac{CI_m}{f_m}$$

$$\frac{\Sigma f_i}{2} = \frac{120}{2} = 60$$

60 lies in the class interval 121–130; hence $LL_m = 121$

$$\text{Median} = 121 + \left[\frac{120}{2} - 30\right] * \frac{10}{31}$$

$$\text{Median} = 130.6$$

As shown in Fig. 6.4, median is a better measure of central tendency as compared to mean for data with outliers or data which is not normally distributed.

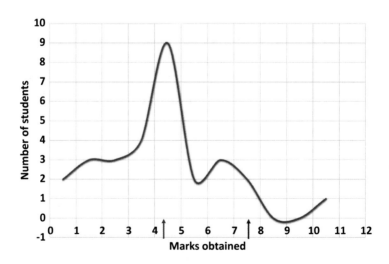

Fig. 6.4 Median (blue arrow) and mean (red arrow) of classified data shown in exercise 6.2

6.3.1   Salient Features of Median

1. Since it is not derived mathematically, it cannot be used in further mathematical calculations.
2. It is an easy and satisfactory measure of central tendency.
3. Since median is the middle value after data arrangement, it is not affected by extreme values (outliers).
4. It is the recommended measure of central tendency for description of data which is not normally distributed.

6.4   Mode

Mode is the most frequently occurring observation in a set of data. The data needs to be classified in the form of a frequency table for observation of the mode. It may or may not be a single observation. For example, if a data has two modes, i.e. two observations with the same frequency, then it is known as bimodal data. The formulas for mode:

$$1. \quad Mode = X_i \qquad \text{(Formula 6.7)}$$

where frequency of ith class (f_i) is maximum among all i

For example (Table 6.5):

Among all serial numbers, f_3 is the maximum of all the frequencies; hence mode is X_3 which is equal to 15.

Exercise 6.7
For the data given in exercise 6.1, calculate mode

Solution

Mean = 55.1
Median = 53
The most frequent occurring observation is 52, Hence, Mode = 52

	i	Observation X_i	Frequency f_i
Table 6.5 frequency distribution of hypothetical data	1.	5	12
	2.	10	6
	3.	**15**	**40**
	4.	20	7
	5.	25	10

Hence, it should be remembered that for normally distributed data, all the three measures of central tendency, i.e. mean, median and mode, are close to each other.

Exercise 6.8

For the data given in exercise 6.2
Mean = 7.76
Median = 5
The most frequent occurring observation is 5, Hence, Mode = 5

2. For classified data

$$\text{Mode} = LL_m + (f_m - f_{m-1}) * \frac{CI_m}{(2f_m - f_{m-1} - f_{m+1})} \qquad \text{(Formula 6.8)}$$

3. Alternate approach for classified data.

$$\text{Let } P = f_m - f_{m-1} \text{ and } Q = f_m - f_{m+1} \text{ then}$$

$$\text{Mode} = LL_m + P * \frac{CI_m}{(P + Q)} \qquad \text{(Formula 6.9)}$$

$$\text{Mode} = LL_m + \frac{CI_m}{\left(1 + \frac{Q}{P}\right)} \qquad \text{(Formula 6.10)}$$

where:

m is the mode class, or the class in which mode lies, i.e. the class which has maximum frequency.
LL_m is the lower limit of the mode class.
f_m is the frequency of the mode class.
f_{m-1} is the frequency of the class preceding the mode class.
f_{m+1} is the frequency of the class succeeding mode class.
CI_m is the class interval of the mode class.

The formula for calculation of mode for classified data can be easily understood by following scenarios (Fig. 6.5):

Scenario I If $P = Q$, i.e. both the class below and the class above have equal frequencies
then, $P/(P + Q) = \frac{1}{2}$, thus

Fig. 6.5 Schematic diagram to explain mode of classified data. In scenario II the mode is being pulled towards the succeeding class, while in scenario III the mode is pulled towards the preceding class

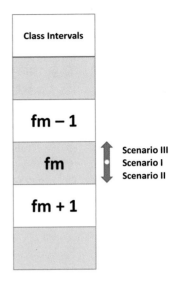

Mode will be exactly in the middle of modal class:

$$\text{Mode} = LL_m + \frac{CI_m}{2}$$

Scenario II if $P > Q$, i.e. the difference between the modal and class preceding the modal class is more, then $P/(P + Q) > \frac{1}{2}$. Mode will lie away from middle of the modal class. It will be more towards the class succeeding the modal class.

Scenario III if $P < Q$, i.e. the difference between the modal class and class preceding the modal class is less, then $P/(P + Q) < \frac{1}{2}$. Mode will lie away from the middle of the modal class. It will be more towards the class preceding the modal class.

6.4.1   Salient Features of Mode

1. Since it is not derived mathematically, it cannot be used in further mathematical calculations.
2. It is an easy and satisfactory measure of central tendency.

Exercise 6.9
For the data given in exercise 6.3, calculate the modal height.

Solution
For calculating modal height using the Formula 6.9.

Table 6.6 Frequency distribution of height

Height in cms	Frequency (f_i)	Cumulative frequency
100–110	11	11
111–120	19	30
121–130	31	61
131–140	21	82
141–150	18	100
151–160	12	112
161–170	8	120
Total	$\Sigma f_i = 120$	

Most frequent class interval is 121–130 (Table 6.6), with $f_i = 31$, which is our modal class

$$P = 31 - 19 = 12$$

$$Q = 31 - 21 = 10$$

$$\text{Mode} = LL_m + P * \frac{CI_m}{(P + Q)}$$

$$\text{Mode} = 121 + 12 * \frac{10}{(12 + 10)}$$

$$\text{Mode} = 126.45$$

Measures of Location

7

Abstract

Background
Percentile
Decile
Quartile

Keywords

Measures of location · Percentile · Decile · Quartile

7.1 Background

Measure of location identifies a value of data at a particular location. Median, which is a measure of central tendency, can also be considered as one of the measures of location because it is the value of data which lies in the middle and divides the data into two equal parts. The measures of location are enumerated in Table 7.1.

Measures of location are frequently used in the clinical medicine, for example, percentiles are incorporated in the growth charts used by paediatricians. Quartile is used as a measure of dispersion (interquartile range; see Chap. 8).

7.2 Percentile

Percentiles are values of a variable which divide the data into 100 equal parts (Fig. 7.1); hence a total of 99 percentiles exist in any given data.

© Springer Nature Singapore Pte Ltd. 2019
S. K. Yadav et al., *Biomedical Statistics*,
https://doi.org/10.1007/978-981-32-9294-9_7

Table 7.1 Measures of location

Measure of location	Division of data (equal parts)
Percentile	100
Decile	10
Quartile	4
Median	2

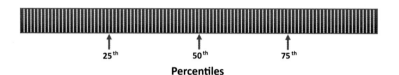

25th   50th   75th

Percentiles

Fig. 7.1   Schematic diagram of percentiles

The value at a particular percentile in a given dataset can be calculated using the following formula:

$$\text{Percentile} = LL_m + \left[\frac{p\Sigma f_i}{100} - C_{m-1}\right] * \frac{CI_m}{f_m} \qquad \text{(Formula 7.1)}$$

where

m is the percentile class, i.e. the class in which the pth percentile lies.
LL_m is the lower limit of the percentile class.
C_{m-1} is the cumulative frequency of the class preceding the percentile class.
f_m is the frequency of the pth percentile class.
CI_m is the class interval of percentile class.

To identify, LL_m location of mth class is derived as follows:

Step 1: Calculate the cumulative frequencies.
Step 2: Calculate l as follows:

$$l = \frac{p\Sigma f_i}{100}$$

Step 3: Find the class in which l lies by looking in the cumulative frequency column.

This concept will be explained in the exercise at the end of this chapter.

7.3   Decile

Deciles are values of a variable which divide the data into 10 equal parts (Fig. 7.2).
Hence a total of nine deciles exist for any given dataset.

The formula to find out the value at a given decile is as follows:

$$\text{Decile} = LL_m + \left[\frac{d\Sigma f_i}{10} - C_{m-1}\right] * \frac{CI_m}{f_m} \qquad \text{(Formula 7.2)}$$

where

m is the decile class or the class in which the dth decile lies.
LL_m is the lower limit of the decile class.
C_{m-1} is the cumulative frequency of the class preceding the decile class.
f_m is the frequency of the dth decile class.
CI_m is the class interval of decile class.

To identify LL_m location of mth class is derived as follows:

Step 1: Calculate the cumulative frequencies.
Step 2: Calculate l as follows:

$$l = \frac{d\Sigma f_i}{10}$$

Step 3: Find the class in which l lies by looking in the cumulative frequency column.

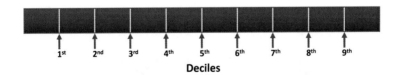

Deciles

Fig. 7.2 Schematic diagram of deciles

7.4   Quartile

Quartiles are values of a variable which divide the data into 4 equal parts (Fig. 7.3). Hence a total of three quartiles exists (Q1, Q2, Q3)

Quartile value can be computed using the following formula:

$$\text{Quartile} = LL_m + \left[\frac{q\Sigma f_i}{4} - C_{m-1} \right] * \frac{CI_m}{f_m} \qquad \text{(Formula 7.3)}$$

where

m is the quartile class, i.e. the class in which the qth quartile lies.
LL_m is the lower limit of the quartile class.
C_{m-1} is the cumulative frequency of the class preceding the quartile class.
f_m is the frequency of the qth quartile class.
CI_m is the class interval of quartile class.

To identify LL_m location of mth class is derived as follows:

Step 1: Calculate the cumulative frequencies.
Step 2: Calculate l as follows:

$$l = \frac{q\Sigma f_i}{4}$$

Step 3: Find the class in which l lies by looking in the cumulative frequency column.

Now, let us understand the concept of the measures of location through the following exercise.

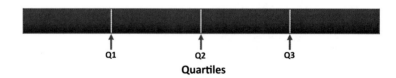

Fig. 7.3 Schematic diagram of quartiles

Exercise 7.1

Calculate the third quartile, eighth decile and 46th percentile from the following data (Table 7.2).

Solution

For third quartile
Using formula 7.3

$$3rd\ Quartile = LL_m + \left[\frac{q\Sigma f_i}{4} - C_{m-1}\right] * \frac{CI_m}{f_m}$$

For location of mth class

$$l = \frac{3 * 42}{4} = 31.5$$

31.5 lies in the class interval 41–50; hence $LL_m = 41$

$$3rd\ Quartile = 41 + \left[\frac{3 * 42}{4} - 21\right] * \frac{10}{14}$$

$$3rd\ Quartile = 48.5$$

For eighth decile
Using formula 7.2

$$Decile = LL_m + \left[\frac{d\Sigma f_i}{10} - C_{m-1}\right] * \frac{CI_m}{f_m}$$

For location of mth class

$$l = \frac{8 * 42}{10} = 33.6$$

33.6 lies in the class interval 41–50; hence $LL_m = 41$

Table 7.2 Frequency and cumulative frequency of data

Class interval	Frequency	Cumulative frequency
11–20	4	4
21–30	7	11
31–40	10	21
41–50	14	35
51–60	7	42

$$8\text{th Decile} = 41 + \left[\frac{8*42}{10} - 21\right] * \frac{10}{14}$$

$$8\text{th Decile} = 50$$

For 46th percentile
Using formula 7.1

$$\text{Percentile} = LL_m + \left[\frac{p\Sigma f_i}{100} - C_{m-1}\right] * \frac{CI_m}{f_m}$$

For location of *m*th class

$$l = \frac{46*42}{100} = 19.32$$

19.32 lies in the class interval 31–40; hence $LL_m = 31$

$$46\text{th Percentile} = 31 + \left[\frac{46*42}{100} - 11\right] * \frac{10}{10}$$

$$46\text{th Percentile} = 39.2$$

Measures of Dispersion

8

Abstract

Background
Range
Interquartile range
Coefficient of quartile deviation
Mean deviation
Mean square deviation (variance)
Variance and standard deviation
Coefficient of variance
Skewness
Kurtosis

Keywords

Mean square deviation · Variance · Standard deviation · Skewness · Kurtosis

8.1 Background

In the results of any study, the values of a variable are not the same, rather they are scattered. This scatter can either be viewed as values within upper and lower range or as values scattered around a mean. The measure of the scatter of data is known as "measure of dispersion" (MoD). The measures of dispersion are enumerated in Table 8.1.

The above-mentioned MoD can be broadly classified as absolute and relative MoD. *Absolute* MoD is expressed in the same statistical unit in which the original data is given, for example, range, quartile deviation and standard deviation. On the other hand, *Relative* MoD denotes a ratio of absolute MoD to an appropriate average. It is also known as *coefficient of dispersion*, for example, coefficient of range, coefficient of quartile deviation and coefficient of variance (CV). There are two

Measures of dispersion
Range
Coefficient of range
Interquartile range
Quartile deviation
Mean deviation
Mean square deviation (variance)
Root mean square deviation (standard deviation)
Coefficient of variance

Table 8.1 Measures of dispersion Range

other measures for description of data which are included in this chapter, namely, *skewness* and *kurtosis*.

8.2   Range

Range, stated simply, is the difference between the largest (L) and smallest (S) value of the data in a data set. It is the simplest measure of dispersion. Range is not based on each and every observation in the data set. A disadvantage of range is that, except for the two extreme values, it does not give any information about the dispersion of the entire data. Hence, dispersion of data from the central tendency cannot be understood from range. In view of these shortcomings, range cannot be further used in mathematical calculations.

$$Range = L - S \qquad\qquad \text{(Formula 8.1)}$$

Coefficient of range is a relative measure of dispersion.

$$\text{Coefficient of range} = \frac{(L - S)}{(L + S)} \qquad\qquad \text{(Formula 8.2)}$$

Exercise 8.1
In a particular study, the height of boys and girls in a class is tabulated below (Table 8.2). Calculate the range of height for boys, girls and the whole class.

Solution

Range $= L - S$.
For boys range is 146 – 178 cm.
For girls range is 127 – 155 cm.
For whole class range is 127 – 178 cm.

Table 8.2 Data of height of boys and girls in a class	Boys (height in cms)	Girls (height in cms)
	150	141
	156	139
	157	127
	160	141
	167	143
	170	135
	146	155
	149	132
	166	137
	161	142
	175	143
	177	139
	164	132
	178	155

8.3   Interquartile Range

Interquartile range, as the name suggests, is the range of values between the first and the third quartiles. This MoD is also not based on each and every observation. Since interquartile range includes quartiles, it is not affected by outliers (extreme values) and, hence, is a more satisfactory MoD than range. However, since this is also a type of range, it cannot be used in further mathematical calculations.

$$\text{Interquartile Range} = Q_3 - Q_1 \qquad \text{(Formula 8.3)}$$

Half of the interquartile range is called semi-quartile range or quartile deviation, given as.

$$\text{Quartile deviation} = \frac{Q_3 - Q_1}{2} \qquad \text{(Formula 8.4)}$$

The interquartile range is used to plot the Box and Whisker plot and to identify the outliers in a data.

Exercise 8.2
Calculate the interquartile range and quartile deviation for the whole class from the data given in exercise 8.1.

Solution

$Q_1 = 140.5$ cm
$Q_2 = 149.5$ cm
$Q_3 = 161.75$ cm

Hence,

Interquartile range $= Q_3 - Q_1 = 161.75 - 140.5 = 21.25$ cm
Quartile deviation $= (Q_3 - Q_1)/2 = (161.75 - 140.5)/2 = 10.625$ cm

8.4   Coefficient of Quartile Deviation

It is a relative measure of dispersion given by the following formula:

$$\text{Coefficient of quartile deviation} = \frac{Q_3 - Q_1}{Q_3 + Q_1} \qquad \text{(Formula 8.5)}$$

8.5   Mean Deviation

The average of absolute deviation of each observation from the mean is known as
mean deviation is given by.

$$\text{Mean deviation} = \frac{\{\Sigma |x_i - \bar{x}|\}}{n} \qquad \text{(Formula 8.6)}$$

Exercise 8.3
Data about number of patients attending out-patient department of a newly opened
hospital in the first 20 days is given below (Table 8.3). Calculate the mean deviation
for this data.

Solution (Table 8.4)

8.6   Mean Square Deviation or Variance

The average of the square of deviation of observations from mean is known as mean
square deviation or variance. It is given by

Table 8.3 OPD attendance in a newly opened hospital

Day	No. of patients
1	30
2	38
3	45
4	55
5	50
6	31
7	45
8	60
9	42
10	39
11	70
12	67
13	84
14	58
15	72
16	87
17	92
18	79
19	59
20	86

Table 8.4 Deviations from mean of the data given in Table 8.3

Day	No. of patients (x_i)	$\bar{x}_i - \bar{x}$	$|x_i - \bar{x}|$
1	30	−29.45	29.45
2	38	−21.45	21.45
3	45	−14.45	14.45
4	55	−4.45	4.45
5	50	−9.45	9.45
6	31	−28.45	28.45
7	45	−14.45	14.45
8	60	0.55	0.55
9	42	−17.45	17.45
10	39	−20.45	20.45
11	70	10.55	10.55
12	67	7.55	7.55
13	84	24.55	24.55
14	58	−1.45	1.45
15	72	12.55	12.55
16	87	27.55	27.55
17	92	32.55	32.55
18	79	19.55	19.55
19	59	−0.45	0.45
20	86	26.55	26.55
Total	1189	$\Sigma_i |x_i - \bar{x}|$	323.9
Mean	59.45	*Mean deviation*	*16.195*

$$\text{Mean square deviation} = \frac{\left\{\Sigma(x_i - \bar{x})^2\right\}}{n - 1} \qquad \text{(Formula 8.7)}$$

Note that standard deviation (s, see below) is nothing but root mean square deviation.

8.7   Variance and Standard Deviation

Variance

$$s^2 = \frac{1}{(n - 1)} \sum (x_i - \bar{x})^2 \qquad \text{(Formula 8.8)}$$

Standard Deviation

$$s = \sqrt{\frac{\sum (x_i - \bar{x})^2}{n}} \qquad \text{(Formula 8.9)}$$

It is also known as root mean square deviation. For smaller samples (less than 20 observation/data points), the formula is as below:

$$s = \sqrt{\frac{\sum (x_i - \bar{x})^2}{n - 1}} \qquad \text{(Formula 8.10)}$$

8.7.1   Concept of Sum of Squares

The entity $\Sigma_i(x_i - \bar{x})^2$ is known as sum of squares. The sum of squares for more than one series has been described in Chap. 13.

$\Sigma_i(x_i - \bar{x})^2$ can be written as

$$\Sigma_i\left(x_i^2 + (\bar{x})^2 - 2\bar{x}x_i\right)$$

$$\Sigma_i x_i^2 + n(\bar{x})^2 - 2\bar{x}\Sigma_i x_i$$

$$\Sigma_i x_i^2 + n\left(\frac{\Sigma_i x_i}{n}\right)^2 - 2\frac{\Sigma_i x_i}{n}\Sigma_i x_i.$$

$$\Sigma_i x_i^2 + \frac{(\Sigma_i x_i)^2}{n} - 2\frac{(\Sigma_i x_i)^2}{n}$$

$$\Sigma_i(x_i - \bar{x})^2 = \Sigma_i x_i^2 - \frac{(\Sigma_i x_i)^2}{n} \qquad \text{(Formula 8.11)}$$

Hence the alternate formula for standard deviation is as follows:

$$s = \sqrt{\frac{\Sigma x_i^2 - \frac{(\Sigma x_i)^2}{n}}{n}} \qquad \text{(Formula 8.12)}$$

8.7.2 Concept of Moment in Statistics

In statistics the entity $\Sigma_i(x_i - \bar{x})^2$ is known as second moment, similarly $\Sigma_i(x_i - \bar{x})^3$ is known as third moment which is utilized in calculation of skewness of data and $\Sigma_i(x_i - \bar{x})^4$ is known as fourth moment which is used in calculation of kurtosis of data.

8.7.3 Features of Standard Deviation

1. The standard deviation is based on each and every observation in the data set.
2. It can be easily and meaningfully used in further mathematical calculations.

From the above formulas, it can be derived that variance can also be written as SD^2. Standard deviation is the most commonly used MoD for quantitative data. Whenever mean is used to describe the central value of a data set, it is flanked by $\pm 2SD$ to denote the dispersion of the same data (Fig. 8.1).

Fig. 8.1 Sample 1 has more dispersion than sample 2

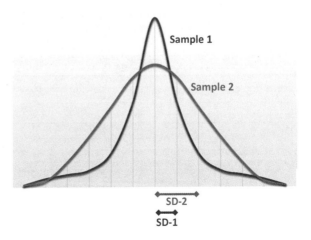

8.8   Coefficient of Variance

Coefficient of variance (CV) is the measure of dispersion which is independent of mean. It is a useful tool for comparison of dispersion of two different variables as it is independent of mean and unit of observation. It is given by:

$$CV = \frac{SD}{\bar{x}} * 100 \qquad \text{(Formula 8.13)}$$

This is frequently used in haematological quality control for assessment of reliability of reagent/method/equipment.

Exercise 8.4

Calculate the mean square deviation (variance), standard deviation and coefficient of variance for the data given in exercise 8.3.

Solution (Table 8.5)

Table 8.5   Square of deviations from mean of the data given in Table 8.3

S. no.	x_i	$x_i - \bar{x}$	$(x_i - \bar{x})^2$
1	30	−29.45	867.303
2	38	−21.45	460.103
3	45	−14.45	208.803
4	55	−4.45	19.803
5	50	−9.45	89.303
6	31	−28.45	809.403
7	45	−14.45	208.803
8	60	0.55	0.303
9	42	−17.45	304.503
10	39	−20.45	418.203
11	70	10.55	111.303
12	67	7.55	57.003
13	84	24.55	602.703
14	58	−1.45	2.103
15	72	12.55	157.503
16	87	27.55	759.003
17	92	32.55	1059.503
18	79	19.55	382.203
19	59	−0.45	0.203
20	86	26.55	704.903
Total	1189	Total	7222.95
Mean	59.45	Mean square deviation $\Sigma_i (x_i - \bar{x})^2$	380.15

$$\text{Variance} = 380.15$$

$$\text{Standard Deviation} = s = \sqrt{\frac{\sum (x_i - \bar{x})^2}{n - 1}}$$

$$s = \sqrt{\frac{7222.95}{20 - 1}}$$

$$s = 19.497$$

Coefficient of variance

$$CV = \frac{SD}{\bar{x}} * 100$$

$$CV = \frac{19.497}{59.45} * 100$$

$$CV = 32.79\%$$

8.9 Skewness

Skewness is a measure of symmetry of data. A data set is called symmetric when it appears similar on both side of the centre point, e.g. normal distribution. Skewness of a normally distributed data is zero. Skewness of a data set is given by Fisher-Pearson coefficient of skewness formula, described below:

$$g_1 = \frac{\sum (x_i - \bar{x})^3/n}{s^3} \qquad \text{(Formula 8.14)}$$

where

$\bar{x}$ is the mean,
s is the standard deviation,
n is the number of data points.
Note that in calculating the skewness, the s is computed with n in the denominator rather than $n-1$.

Fig. 8.2 Concept of negative
and positive skewness

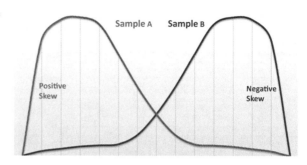

8.9.1   Features of Skewness

1. Skewness of a normal distribution curve is zero.
2. Negative value of skewness denotes that the data is skewed towards the left side, i.e. the left tail of the curve is longer than the right tail (Fig. 8.2).
3. Positive value of skewness denotes that the data is skewed towards the right side, i.e. the right tail of the curve is longer than the left tail (Fig. 8.2).

8.10   Kurtosis

Kurtosis is a measure of whether the data is heavy-tailed or light-tailed relative to a normal distribution. This means that data sets with positive kurtosis (leptokurtic) tend to have more outliers. Data sets with negative kurtosis (platykurtic) tend to have less outliers. The formula for kurtosis is as follows:

$$\text{Kurtosis} = \frac{\sum (x_i - \bar{x})^4/n}{s^4} - 3 \qquad \text{(Formula 8.15)}$$

where

$\bar{x}$ is the mean,
s is the standard deviation,
n is the number of data points.
Note that in calculating the Kurtosis, the s is computed with n in the denominator rather than $n-1$.
The kurtosis for a standard normal distribution is zero (mesokurtic).

Exercise 8.5
In a particular study involving 175 college-going students, the values for haemoglobin level were determined. The data is depicted as frequency table given

below (Table 8.6). Plot the data on a graph, and find out skewness and kurtosis for the given data.

Solution (Table 8.7)
For the given data, a frequency polygon can be drawn as shown in Fig. 8.3.

$$\text{Skewness} = \frac{\sum (x_i - \bar{x})^3 / N}{s^3}$$

$$\text{Skewness} = \frac{-263.624 / 175}{2.57^3}$$

$$\text{Skewness} = -0.089$$

Since the skewness is negative, the graph is expected to be left skewed, as evident in Fig. 8.3

Kurtosis

$$\text{Kurtosis} = \frac{\sum (x_i - \bar{x})^4 / N}{s^4} - 3$$

Table 8.6 Haemoglobin values of 175 college students

Haemoglobin (g/dl)	Number of patients
4-6	5
7-9	18
10-12	80
13-15	60
16-18	12

Table 8.7 Table for calculation of skewness and kurtosis

Haemoglobin	Frequency	Mid point (x_i)	$f_i x_i$	$x_i - \bar{x}$	$(x_i - \bar{x})^2$	$(x_i - \bar{x})^3$	$(x_i - \bar{x})^4$
4–6	5	5	25	−6.96	48.44	−337.15	2346.59
7–9	18	8	144	−3.96	15.68	−62.10	245.91
10–12	80	11	880	−0.96	0.92	−0.88	0.85
13–15	60	14	840	2.04	4.16	8.49	17.32
16–18	12	17	204	5.04	25.40	128.02	645.24
Total=	175		2093		94.61	−263.62	3255.91
	$\bar{x} = 11.96$			$s = 2.57$			

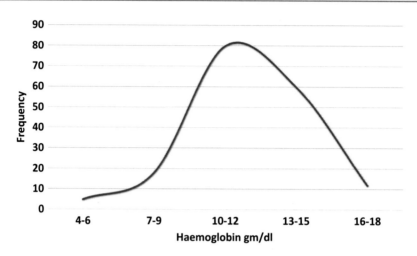

Fig. 8.3 Frequency polygon of the data given in Table 8.6

$$\text{Kurtosis} = \frac{3255.911/175}{2.57^4} - 3$$

$$\text{Kurtosis} = -2.574$$

Negative value of kurtosis suggests that the data set is platykurtic.

Sampling Methods

9

Abstract

Probability sampling
  Simple random sampling
  Simple systematic sampling
  Stratified random sampling
  Stratified systematic sampling
  Multistage sampling
  Multiphase sampling
  Cluster sampling
  Event sampling
Non-probability sampling
  Convenience sample or accidental sampling
  Purposive sample or judgemental sampling
  Quota sampling
  Snowball sampling

Keywords

Probability sampling · Non-probability sampling · Simple random · Stratified
systematic · Cluster sampling · Event sampling

9.1 What Is Sampling?

When each and every subject/individual/item/object of the population is assessed or
measured, the process is known as *census*. On the other hand, when information is
collected from a portion of the population, it is known as *sampling*. Sampling is
required because census is costly, time-consuming and labour intensive. Moreover,
sometimes the entire population may not be accessible for the assessment of a

© Springer Nature Singapore Pte Ltd. 2019   71
S. K. Yadav et al., *Biomedical Statistics*,
https://doi.org/10.1007/978-981-32-9294-9_9

variable. As we have learnt earlier, sample represents the population from which it is derived, and the results from the study on this sample are extrapolated to that population. Hence, it is of paramount importance that the sample is as representative of the population as possible. Factors which affect representativeness of a sample include sampling method, sample size and participation (especially when involving human subjects). Hence, the methods to derive the sample from a population need to be carefully considered. There are a number of methods of sampling that we should apprise ourselves of, before undertaking any epidemiological study. Sampling frame is the list of subjects/individuals/items/objects from which a sample has to be derived (Fig. 9.1).

Sampling methods can be broadly classified into 2 types:

9.1.1   Probability Sampling

In probability sampling, each and every unit of population has a chance (the chance may or may not be equal) of selection into the sample, i.e. the probability of selection of each unit is more than zero. When there is an equal chance of selection, it is

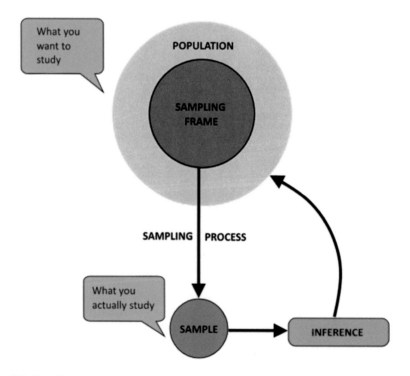

Fig. 9.1 Sampling process

known as equal probability of selection (EPS) sample design. The advantage of probability sampling is that since the probability of selection is known, the precision/measure of uncertainty of data can be accurately estimated.

9.1.2 Non-probability Sampling

Unlike the probability sampling method, non-probability sampling does not provide chance of selection to some of the units of population. In this case, probability of selection of a unit cannot be accurately determined, and hence "sampling errors" cannot be estimated. Due to the design of sampling, non-probability sampling methods are liable to yield biased results.

The various sampling methods available to the researchers are tabulated in Table 9.1 and have been described individually later in this chapter.

9.2 Sampling Process

In brief, the sampling process comprises of the following steps

1. Defining the population of interest
2. Specifying a sampling frame within the population, i.e. a set of items or events possible to measure
3. Specifying a sampling method for selecting items or events from the sampling frame
4. Determining the sample size using one of the methods described later in this book (Chap. 18)
5. Implementing the sampling plan
6. Sampling and data collection
7. Reviewing the sampling process for appropriateness of the collected data

Table 9.1 Probability and non-probability sampling methods

Probability sampling	Non-probability sampling
Simple random sample	Convenience sample or accidental sampling
Simple systematic sample	Purposive sample or judgemental sampling
Stratified random sample	Snowball sampling
Stratified systematic sample	Quota sampling
Multistage sample	
Multiphase sample	
Cluster sample	

9.3   Without Replacement (WOR) vs with Replacement (WR)

Sampling schemes may be without replacement ("WOR" – no element can be selected more than once in the same sample) or with replacement ("WR" – an element may appear multiple times in the sample). In biomedical statistics, mostly without replacement sampling is done, so that none of the individual patients is included more than once in the sample.

We shall now learn about the various sampling methods with illustrations using the homogeneity or heterogeneity of the population as a basis.

9.4   Probability Sampling

There are a number of methods for probability sampling, each having its own merits and fallacies. The selection of the method depends on the characteristics of the population as described in the situations in the following sections.

9.4.1   Situation-Homogenous Population: Method-1, Simple Random Sampling

In simple random sampling (Fig. 9.2), the desired number of subjects are selected randomly using the lottery method or by the use of random number tables. This method of sampling is possible only when a complete list of individuals constituting the population is available.

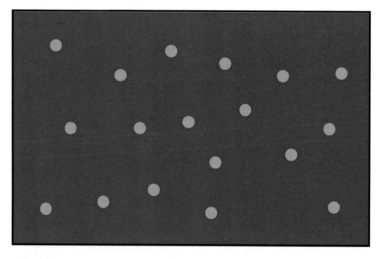

Fig. 9.2 Sampling for homogenous population – random method (rectangle represents the population, and dots are the individual samples)

9.4.1.1 Advantages
- Estimates are easy to calculate.
- Simple random sampling is always an EPS design, but not all EPS designs are simple random sampling.

9.4.1.2 Disadvantages
- This method is impractical if the sampling frame is too large.
- Minority subgroups of interest in the population may not be present in the sample in sufficient numbers for study. In such a scenario, other methods of sampling should be applied.

The simple random sampling method is illustrated in Fig. 9.3 to explain the unequal gaps in the sample.

9.4.2   Situation-Homogenous Population: Method-2, Simple Systematic Sampling

Starting from a random point on a sampling frame (selected by lottery or random number table), every nth element in the frame is selected at equal intervals (sampling interval). In this method, the first object is selected at random. This is followed by selection of objects at equal suitable interval to sample the entire population (Figs. 9.4 and 9.5). The sampling interval can be obtained by dividing the total population size by the desired sample size. For this sampling method also, a complete list of individuals must be available.

9.4.2.1 The Periodicity Problem of Simple Systematic Sampling
If the periodicity calculated for the sample matches, the periodicity of occurrence of a phenomenon in the population under study (phenomenon m Fig. 9.6), then the sample is no longer random. In fact, it may be grossly biased towards either inclusion or exclusion of such occurrences. So, which type of sampling is more appropriate in this situation? The answer is simple random sampling, since the biasness due to periodicity of the occurrence of a phenomenon may be avoided by simple random sampling method.

The gap, or period between successive elements is random, uneven and has no particular pattern

Fig. 9.3 Scheme of simple random sampling

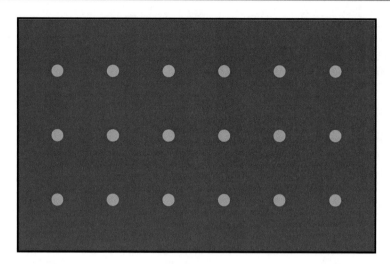

Fig. 9.4 Sampling for homogenous population – systematic method (rectangle denotes the population, and dots represent the individual sample)

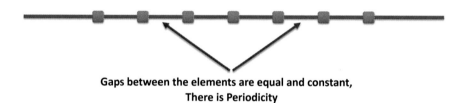

Gaps between the elements are equal and constant,
There is Periodicity

Fig. 9.5 Scheme of simple systematic sampling

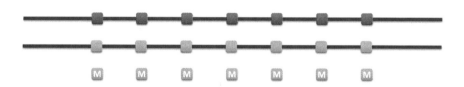

Fig. 9.6 Depiction of the periodicity problem

9.4.2.2 Advantages
- Sample is easy to select since only the first object has to be identified randomly followed by periodic sampling.
- Suitable sampling frame can be identified easily
- Sample is evenly spread over the entire reference population

9.4.2.3 Disadvantages

- Sample may be biased if the hidden periodicity in population coincides with that of selection, as shown in the Fig. 9.6.
- It may be difficult to assess the precision of estimate from one survey. Multiple surveys may be required with different starting point in each survey to be able to confidently assess the precision.

9.4.3 Situation-Heterogeneous Population: Stratified Random/ Systematic Sampling

Stratified random/systematic sampling is used for populations with multiple strata where elements within each stratum are homogeneous but are heterogeneous across the strata. In such a situation, the entire population is first divided into subgroups, depending upon its heterogeneity, so that the individual sub-groups or strata become homogenous. Then from each thus created subgroup or strata, the samples are drawn at random or in a systematic manner, as described earlier, in proportion to its size (Fig. 9.7). Since each stratum is treated as an independent population, different sampling approaches can be applied to different strata also.

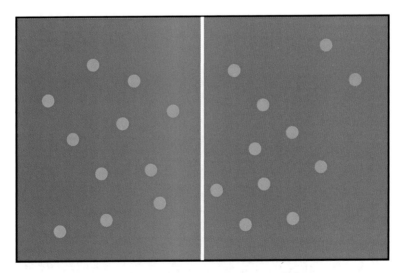

Fig. 9.7 Sampling for heterogeneous population – random method (rectangle represents the population, and dots are the individual sample)

9.4.3.1 Disadvantages

- For division of the population into strata, the sampling frame of entire population has to be prepared separately for each stratum.
- When examining multiple criteria in the same population, stratifying variables may be related to some, but not to other criteria. This complicates the sampling design and potentially reduces the utility of creating strata.
- In some cases, such as designs with a large number of strata, or those with a specified minimum sample size per group, stratified sampling can potentially require a larger sample than other sampling methods.

9.4.4   Situation-Heterogeneous Population with Clusters: Cluster Sampling

A population can have pre-existing/natural groups which are heterogeneous internally but homogeneous to all the other clusters. A cluster is usually a predefined/natural group in a population. In cluster sampling, the clusters are first chosen in a random fashion or in systematic manner as described earlier (Fig. 9.8). Subsequently, all subjects from the chosen clusters are taken for data collection. Cluster sampling is often used to evaluate vaccination coverage in EPI (extended programme of immunization by Government of India).

Note that clusters are internally heterogeneous, whereas groups/subgroups in the sampling methods described earlier are internally homogenous.

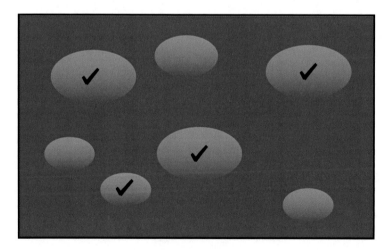

Fig. 9.8 Sampling for heterogeneous population with clusters – cluster sampling (rectangle represents the population, and ovals are the clusters)

9.4.4.1 Advantages
- This method of sampling cuts down on the cost of preparing a sampling frame.
- This method helps to reduce travel and other administrative costs in evaluation of implementation of national programmes.

9.4.4.2 Disadvantages
- Sampling error in cluster sampling is higher than a simple random sample of the same size from the same population.

Difference Between Strata and Clusters
Although strata and clusters both are nonoverlapping subsets of the population, they differ in several ways:

- All strata are represented in the sample, whereas only a subset of clusters is included in the sample.
- With stratified sampling, the best survey results occur when elements within strata are internally homogeneous. However, with cluster sampling, the best results occur when elements within clusters are internally heterogeneous

9.4.5 Situation-Heterogeneous Population with Groups and Subgroups: Multistage Sampling

This is used for heterogeneous population with predefined groups, subgroups, sub-subgroups and so on. In multistage sampling, sample is collected over several stages. First a sample of groups is taken. Then from the selected groups, subgroups are chosen, and so on till the last subgroup classification of the population is reached. Within each group or subgroup, the sampling can again be random or systematic (Fig. 9.9).

9.4.6 Situation-Multiple Parameters with Selective Collection: Multiphase Sampling

In this situation also, the population has groups and subgroups. However, in contrast to the multistage sampling, this sampling process collects the sample in multiple phases. This means that some information is collected from the whole population followed by another information from a relevant group and so on. Survey using such a procedure is less costly, less laborious and more purposeful of the information to be collected and analysed.
  For example:

- In survey of prevalence of tuberculosis, Mantoux test (MT) is done in all cases – Phase I
- X-ray chest only in MT + ve cases – Phase II
- Sputum examination in X-ray +ve cases – Phase III

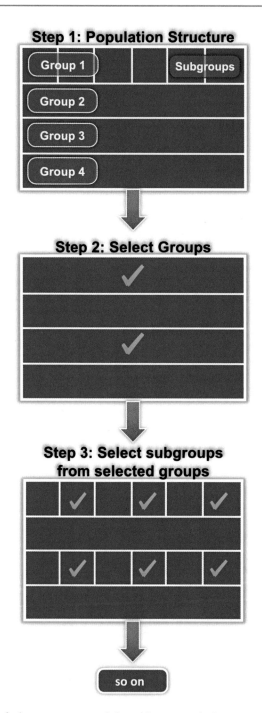

Fig. 9.9 Sampling for heterogeneous population with groups and subgroups – multistage sampling

9.4.7   Event Sampling

Event sampling methodology (ESM) is a new form of sampling method that allows researchers to study the ongoing experiences and events that vary within as well as across days in the naturally occurring environment of study participants. Given the frequent sampling of events inherent in ESM, this method enables researchers to measure the typology of activity and detect the temporal and dynamic fluctuations of work experiences.

For example, a study of job description of a particular job may be conducted using the ESM method, considering that the job description has a potential to change over time. Similarly, if eating is considered as an event, the detailed food habits of an individual can be sampled using ESM.

There are three types of ESM: signal contingent, event contingent and interval contingent depending on whether the trigger is a predefined signal, event or a specified time interval, respectively.

9.4.7.1 Disadvantage

Since this requires frequent event sampling in the naturally occurring environment, it can be perceived as intrusive by some of the participants. Moreover, ESM may substantially change the phenomenon being studied, since the sampled participant may modify his/her behaviour in response to the triggers knowing the fact that he/she is under study.

To summarize the situational utility of probability sampling, refer to Table 9.2.

9.5   Non-probability Sampling

9.5.1   Convenient Sampling

It is a type of non-probability sampling also known as *accidental sampling*. In this sampling method, the selection of units of population is at the convenience of the investigator (Fig. 9.10), e.g. data of exit polls collected by a press reporter outside a polling station. It has all the disadvantages of non-probability sampling described

Table 9.2 Sampling methods corresponding to various situations

Situations	Sampling methods
Homogenous population	Simple random sample
	Simple systematic sample
Heterogeneous population	Stratified random sample
	Stratified systematic sample
Heterogeneous population with clusters	Cluster sample
Heterogeneous population with groups/subgroups	Multistage sample
Special circumstances	Multiphase sample
	Event sampling

Fig. 9.10 Convenient
sampling: Investigator
including only proximal
individuals as per his
convenience in the study

earlier like biasness of results and inability to determine sampling errors. However, convenient sampling is quite useful in exploratory study designs and generation of hypothesis, and, hence, it is frequently used in many hospital-based studies.

9.5.2   Judgemental Sampling

It is a type of non-probability sampling also known as *purposive sampling*. Here, those units of population are selected by the investigator, which according to him/her is representative of the population.

9.5.3   Quota Sampling

It is a type of non-probability sampling, where the units of population are selected on the basis of features like age, gender, income, etc. which have predefined "quotas".

9.5.4   Snowball Sampling

This type of non-probability sampling is also known as *chain-referral sampling*. In snowball sampling method, the investigator selects the first few samples and then either recruits them or asks them to recommend other subjects they know who would fit the description of samples needed. It can be either linear, exponential non-discriminative and exponential discriminative sampling (Fig. 9.11).

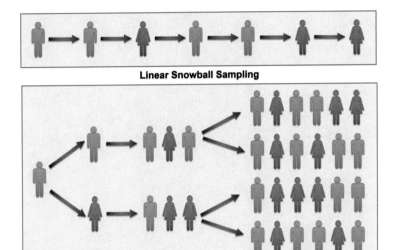

Linear Snowball Sampling

Exponential Non-discriminative Snowball Sampling

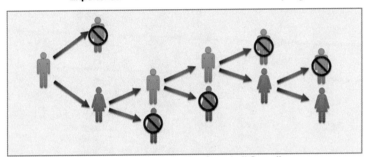

Exponential Discriminative Snowball Sampling

Fig. 9.11 Snowball sampling

Part II

Tests for Inference

Statistical Distribution-Continuous

10

Abstract

Background
Normal distribution
Standard normal distribution
Student t-distribution
Chi-square distribution
F-distribution

Keywords

Normal distribution · Standard normal deviate · t-distribution · Chi-square distribution · F-distribution

10.1 Background

Since the application of sampling methods depend on the distribution of the variable of interest in the population, let us move ahead and understand the concept of this distribution. The probability of occurrence of a particular value in a given data set usually follows a particular pattern, and this pattern is known as the "statistical distribution".

Frequency polygon and probability distribution (graph) are similar. In a probability distribution, the frequencies are converted to probabilities by summing up the total frequencies and dividing each individual frequency by the total. The depiction of these calculated probabilities is known as the "probability density function" (PDF, see Chap. 26). On the other hand, if the probabilities are depicted as cumulative ones and a curve is plotted, the same is called as "cumulative density function" (CDF, see Chap. 26). Remember that the frequency polygon and probability density function are not the same. Sum of all the frequencies in a frequency distribution is equal to the number of cases, while the probabilities in a probability density function add up to one, since probabilities can only be between 0 and 1.

© Springer Nature Singapore Pte Ltd. 2019
S. K. Yadav et al., *Biomedical Statistics*,
https://doi.org/10.1007/978-981-32-9294-9_10

The distribution of a variable in consideration may be either "continuous" or "discrete". Discrete statistical distribution is described later in Chap. 27 of this book. In the present chapter, we will understand the concept of a "continuous" statistical distribution. Examples of the possible statistical distribution of a continuous variable include:

(a) Normal distribution.
(b) Standard normal distribution.
(c) Student t-distribution.
(d) Chi-square distribution.
(e) F-distribution.

Let us briefly understand the concept of each of them with examples.

10.2   Normal Distribution

In normal distribution, PDF curve of data is bell-shaped (Fig. 10.1). The mean of such a data lies at the peak of this curve. Multiples of standard deviation (SD) are considered to include certain percentages of the data, as shown in Table 10.1.

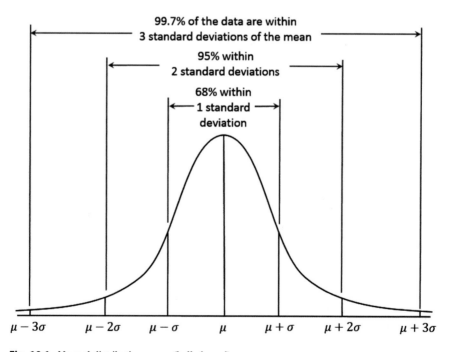

Fig. 10.1 Normal distribution curve (bell-shaped)

Table 10.1 Dispersion of normal distribution	Distance from mean	Percentage of cases covered
	Mean ± 1 SD	68.27%
	Mean ± 2 SD	95.45%
	Mean ± 3 SD	99.73%

Given the bell shape of the curve, the measures of central tendency, i.e. mean, median and mode, of a normal distribution curve coincide. The graph is symmetric about its mean, and the skewness is zero. It is noteworthy that if a population follows a normal distribution, the "distribution of means" of multiple samples taken from a population will also follow a normal distribution. Similarly, in such a situation "difference between two means" will also follow a normal distribution.

10.2.1 Test for Normality

If skewness of data (Formula 8.14) indicates that data is symmetrical and kurtosis (Formula 8.15) indicates that data is mesokurtic, then it can be assumed that the data is normally distributed. However, there are specialized tests available, namely, Shapiro-Wilk test and Kolmogorov-Smirnov normality test, which can be used to test the normality of data.

10.3   Standard Normal Distribution

This is a special type of normal distribution with a mean of "0" and standard deviation of "1". Since this is also normal distribution, similar amount of data is bound together by multiples of SD as shown for a normal distribution (Table 10.1).

10.3.1 Utility of Standard Normal Distribution

It is used for the following purposes:

1. To find the percentage of cases beyond either side of a particular value within the sample (see exercise 10.2).
2. To find a particular value within the sample where a fixed number of cases are less or more than that value (see exercise 10.3).
3. To find the percentage of cases between two selected values.
4. For inference testing based on z value.

10.3.2 Standard Normal Deviate

It is the deviation of an observation from mean in terms of standard deviation for a normally distributed data. It is given by the following formula:

$$z_i = \frac{(x_i - \bar{x})}{SD} \qquad \text{(Formula 10.1)}$$

where

z_i is the standard normal deviate for x_i, the ith value.

The distribution of standard normal deviate is nothing but a standard normal distribution. Thereby the mean of all z_i will be "0" and standard deviation of all z_i will be "1".

10.3.3 Concept of Centred Data and Standardized Data

Centring the Data
If each value of data is subtracted from the mean of the data, the set of resulting values is centred data:

$$c_i = (x_i - \bar{x})$$

The mean of the resulting data becomes 0.

Standardizing the Data
If values of data are converted into z score, then the resulting data set is a standardized data.

$$z_i = \frac{(x_i - \bar{x})}{SD}$$

The mean of the resulting data is 0, with standard deviation of 1.
The concept of normal distribution will become clearer with the following exercises:

Exercise 10.1
The average weight of babies at birth in a population is 3 kg with SD of 0.25 kg. If the birth weights are normally distributed, would you regard:

(a) Weight of 4 kg as abnormal?
(b) Weight of 2.56 kg as normal, assuming that values outside two SD are outliers?

Solution

(a) z_i for weight of 4 kg is equal to $(4-3)/0.25 = 4$. Since $z_i > 1.96$, 4 kg weight can be considered as abnormal.
(b) z_i for weight of 2.56 kg is equal to $(2.56-3)/0.25 = -1.76$. Since $z_i < 1.96$, a weight of 2.56 kg can be considered as normal.

Exercise 10.2
From the data given in exercise 10.1, identify how many children will have birth weight of more than 3.25 kg.

Solution
z_i for weight of 3.25 kg is equal to $(3.25–3)/0.25 = 1$, $p = 0.1587$. Hence, about 15.87% subjects will be more than 3.25 kg.

Exercise 10.3
From the data given in exercise 10.1, identify the birth weight beyond which 2.5% children lie in the distribution.

Solution
For 2.5%, i.e. for $p = 0.025$, Z_i will be 1.96. Hence, from

$$\frac{(x_i - 3)}{0.25} = 1.96$$

x_i will be 3.49. So, 2.5% of the babies will have birth weight of more than 3.49 kg.

10.4   Student t-Distribution

Student t-distribution is a statistical distribution published by William Gosset in 1908. Ironically, his employer, Guinness Breweries, required him to publish under a pseudonym rather than his actual name, so he chose "Student", and concepts given by him carry the name Student. In this distribution, if df is degree of freedom, then mean $= 0$, variance $= df/(df–2)$, kurtosis $= 6/(df–4)$ and skewness $= 0$ (Fig. 10.2).
    The shape of t-distribution is determined by its degree of freedom. For larger degree of freedom, the distribution becomes similar to normal distribution (Fig. 10.3).

10.4.1 Applications

The applications of Student t-distribution are the same as those of standard normal distribution but with a smaller sample size. Hence, this distribution can be used for either of the following:

Fig. 10.2 Shape of
t-distribution

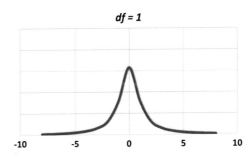

df = 1

-10   -5   0   5   10

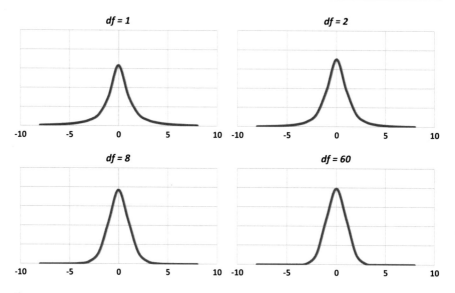

Fig. 10.3 Shape of t-distribution depends on sample size (size increases from top to down and left to right)

1. To find percentage of cases beyond either side of a particular value for a smaller sample size (less than 30).
2. To find particular value with a fixed number of cases lesser than or more than that value for a smaller sample size.
3. For testing for hypothesis (comparison of means) for smaller sample size.
4. For testing of correlation coefficient for smaller sample size.

10.5   Chi-Square Distribution

Like the Student t-distribution, the shape of chi-square distribution is also determined by its degrees of freedom. This distribution is skewed towards the right, especially at lower degrees of freedom (Fig. 10.4).

10.5.1 Applications

The chi-square distribution is used for testing of hypothesis involving nominal or ordinal data, specifically:

1. Measure of association between two variables.
2. Measure of "goodness of fit" of a probability function or model.

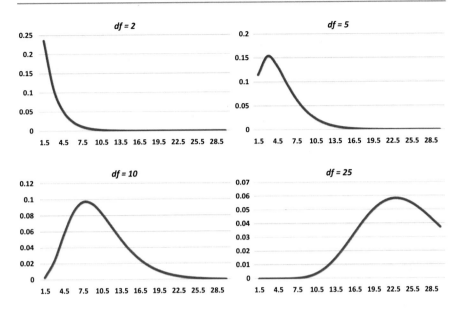

Fig. 10.4 Shape of chi-square distribution depends on degree of freedom

10.6   F-Distribution

The last type of continuous distribution having biomedical importance is the F-distribution. This is a right-skewed distribution and is used in Analysis of Variance (ANOVA/MANOVA). The F-distribution is actually a ratio of two chi-square distributions, and a specific F-distribution is characterized by the degrees of freedom for the numerator ($df1$) chi-square and the degrees of freedom for the denominator ($df2$) chi-square (Fig. 10.5).

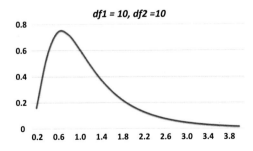

Fig. 10.5 Shape of F-distribution for degree of freedom 10,10

Sampling Distribution and Hypothesis Testing

11

Abstract

Background
Confidence interval
Difference in means
Concept of z-statistics
    Concept of one-tailed and two-tailed test
    Critical region on either side of means
Hypothesis Testing

Keywords

Confidence interval · Hypothesis testing · One-tailed · Two-tailed · Power of study · Type I error

11.1 Background

Now that we have understood the basics of statistical distribution and sampling methods, we can move on to understand the concept of hypothesis testing which is the main application of biomedical statistics. For this, we introduce the concept of standard error (SE). The means of multiple samples (1 to k) from a population will follow normal distribution and have a variance denoted by square of standard error (SE^2).

$$\text{Sampling distribution variance} = \frac{SD^2}{n} \qquad \text{(Formula 11.1)}$$

where $\frac{SD}{\sqrt{n}}$ is known as standard error. Using SE, the confidence interval of mean is calculated.

11.2 Confidence Interval

For a population with normal distribution, about 95% of means (of sampling distribution) will lie within $\mu \pm 1.96SE$. This statement can be extended to saying that, in 95% of instances, the population mean will lie within $\bar{x} \pm 1.96$ SE. This interval is known as the confidence interval.

$$95\% CI = \bar{x} \pm 1.96\ SE \qquad\qquad \text{(Formula 11.2)}$$

Exercise 11.1
Average weight of 100 babies at birth is 3.0 kg with the SD of 0.25 kg. If the birth weights are normally distributed, what is the 95% confidence interval for this sample mean?

Solution

$$SE = 0.25/\sqrt{100} = 0.025.$$

$$95\%\text{CI} = 3 \pm 1.96 * 0.025 = 2.951 \text{ to } 3.049\ kg$$

11.3 Difference in Means

Similarly, *difference in means* of a population with normal distribution also follows a normal distribution, as alluded to earlier with a variance given by

$$s^2\left(\frac{1}{n_1} + \frac{1}{n_2}\right) \qquad\qquad \text{(Formula 11.3)}$$

where

$$s^2 = \frac{n_1 s_1^2 + n_2 s_2^2}{n_1 + n_2}$$

where $\left(n_1 s_1^2 + n_2 s_2^2\right)$ is nothing but pooled "sum of squares" of two samples.

11.4 Concept of z-Statistics

Extending the concept of standard normal deviate explained in Chap. 10, we can calculate statistics z_a as

$$z_a = \frac{a - E(a)}{\sqrt{V(a)}} \qquad \text{(Formula 11.4)}$$

where $E(a)$ is expectation of variable (a) and $V(a)$ is variance of variable (a).
  Hence, z_a will follow standard normal distribution.

11.4.1 Concept of One-Tailed and Two-Tailed Test

This is an important concept since it is applicable to most of the statistical tests. If we want to know that statistics of the test group is more than (or less than) the control group, then it is known as one-tailed test.

  If we want to know that statistics of the test group is different (it may be more than or less than) the control group, then it is known as two-tailed test. We shall be applying this concept in further chapters.

11.4.2 Critical Region on Either Side of Means

Critical region is that part of the curve (with corresponding area under the curve), where if our value falls, it would result in rejection of the null hypothesis (see below). For two-tailed test, if any value falls in the $\alpha/2$ critical region on either side, we reject the null hypothesis. For one-tailed test, if any value falls in the α critical region on appropriate side, we reject the null hypothesis. This shall become clearer once we understand the concept of null hypothesis, described next.

11.5 Hypothesis Testing

Our belief of what is true (or what we expect to be correct) is known as "hypothesis". It is denoted by H_1 and is also known as "alternate hypothesis". Opposite of the alternate hypothesis is known as the "null hypothesis" which is denoted by H_0. We undertake a study and its statistical analysis to either accept or reject the null hypothesis.
  For example:

Hypothesis: Alcohol causes liver cell injury.
Alternate hypothesis (H_1): Alcohol causes liver cell injury.
Null hypothesis (H_0): Alcohol does not cause liver cell injury.

  In such a situation following possibilities may arise as tabulated in Table 11.1.

Table 11.1 Concept of type I and type II error

		True situation	
		H0	H1
Inference	H0	Acceptance of null hypothesis	Type II error (β)
	H1	Type I error (α)	Rejection of null hypothesis

11.5.1 Type I Error

As can be seen from Table 11.1, type I error occurs when the null hypothesis is rejected while being true. It is also known as alpha (α). The usual cut-off for type I error in medical studies is 0.05 or 5%. Hence $\alpha/2$ is 0.025 or 2.5%.

11.5.2 Type II Error

Acceptance of null hypothesis while being false is called as type II error. This is also known as beta (β). The usual cut-off for type II error is considered at 0.2 or 20%.

11.5.3 Power of Study

This is the probability of rejecting the null hypothesis when it is false. Mathematically it is $1-\beta$ and hence is commonly kept at 80%. Power can be estimated before the study is undertaken (a priori) or after the study has been completed (post hoc).

It is obviously desirable to keep both the type I error (α) and type II error (β) as low as possible for the inference to be close to the real situation. In medical science, type I error is more hazardous than type II error. This is so because rejection of an otherwise true *null hypothesis* may lead to acceptance of spurious association between two factors of a cause of disease. In drug trials, type I error is more dangerous because higher efficacy of one drug over another may be accepted while it was not true. Hence, studies need to be designed in such a way that type I error is kept at minimum. Appropriate increase in sample size can reduce both errors , and hence, accurate calculation of the sample size and adherence to the same are desirable.

Test of Inference: One-Sample or Two-Sample Mean

12

Abstract

Background
One sample mean
Two sample mean
Test for homogeneity of variance
Two small sized sample- Test for inference
Test for inference of paired samples

Keywords

One-sample mean · Two-sample mean · Homogeneity of variance · Paired t-test · Student t-test

12.1 Background

Now, let us move on to the various statistical tests for inference one by one. It is essential to understand the application of these tests in terms of when to apply and under what assumptions to apply. It is also important to frame the null hypothesis judiciously.

When we draw sample from a population, it is prudent to know whether the sample mean differs from the population mean.

12.2 One-Sample Mean (z-test)

If a study is done with a sample drawn from a population with mean (μ), then by using the z-test we can assess the significance of sample mean, i.e. whether the sample is significantly different from population means.

(a) For large sample z is given as

$$z = \frac{\bar{x} - \mu}{\sqrt{\frac{\sigma^2}{n}}} \qquad \text{(Formula 12.1)}$$

(b) For small sample

$$z = \frac{\bar{x} - \mu}{\sqrt{\frac{\sigma^2}{n-1}}} \qquad \text{(Formula 12.2)}$$

where $\bar{x}$ is sample mean, μ is population mean and σ is sample standard deviation.

In the above formulae, it is assumed that the data is normally distributed within the population as well as the sample drawn thereof.

Exercise 12.1

The mean haemoglobin of a female population is 11.2 g/dl. Haemoglobin of a sample of 80 females from a population was measured. Mean haemoglobin of the sample was 12.5 g/dl with an SD of 2.5. Determine whether the sample mean is different from the population mean.

Solution

Using formula

$$\mu = 11.2$$

$$\bar{x} = 12.5$$

$$n = 80$$

$$\sigma = 2.5$$

$$z = \frac{\bar{x} - \mu}{\sqrt{\frac{\sigma^2}{n.}}}$$

$$z = \frac{12.5 - 11.2}{\sqrt{\frac{2.5^2}{80}}}$$

$$z = 4.66$$

Since calculated z is more than 1.96, the sample mean is significantly different from the population mean.

12.3   Two-Sample Mean (z-test)

If two samples are drawn, then it can be assessed whether they belong to the same sampled population by using z-test. As has been described previously, difference between two means also follows normal distribution with a variance of

$$s^2\left(\frac{1}{n_1} + \frac{1}{n_2}\right)$$

Hence, z-statistics for inference is given as

$$z = \frac{(\bar{x}_1 - \bar{x}_2) - 0}{\sqrt{s^2\left(\frac{1}{n_1} + \frac{1}{n_2}\right)}} \qquad \text{(Formula 12.3)}$$

where $s^2 = \frac{n_1 s_1^2 + n_2 s_2^2}{n_1 + n_2}$.

A shortcut approximation of above formula is as follows:

$$z = \frac{(\bar{x}_1 - \bar{x}_2) - 0}{\sqrt{\frac{s_1^2}{n_1} + \frac{s_2^2}{n_2}}} \qquad \text{(Formula 12.4)}$$

In the above formula too, it has been assumed that the sample size of both samples is large enough to calculate the variance and that the data is normally distributed.

Exercise 12.2

In a particular study, the random blood sugar level of 300 males and other group of 400 males was measured. First group had a mean blood sugar level of 95 mg/dl and SD of 3.5, while other group had a mean of 80.5 mg/dl and SD of 2.9. Determine if the difference between the two groups is statistically significant.

Solution

$$n_1 = 300$$

$$n_2 = 400$$

$$\bar{x}_1 = 95$$

$$s_1 = 3.5$$

$$\bar{x}_2 = 80.5$$

$$s_2 = 2.9$$

$$s^2 = \frac{n_1 s_1^2 + n_2 s_2^2}{n_1 + n_2}$$

$$s^2 = \frac{300 * 3.5^2 + 400 * 2.9^2}{300 + 400}$$

$$s^2 = 10.05$$

$$z = \frac{(\bar{x}_1 - \bar{x}_2) - 0}{\sqrt{s^2 \left(\frac{1}{n_1} + \frac{1}{n_2}\right)}}$$

$$z = \frac{(95 - 80.5) - 0}{\sqrt{10.05 \left(\frac{1}{300} + \frac{1}{400}\right)}}$$

$$z = 4.57$$

Since calculated z is more than 1.96, the two-sample means are significantly different.

12.4   Test for Homogeneity of Variance in Two Samples (F-test)

When comparing two samples, it is also important to assess whether the two data sets have equal variance or not. This is done by F-test by calculating the following ratio:

$$F \text{ ratio} = \frac{\text{Var } 1_{high}}{\text{Var } 2_{low}} \qquad \text{(Formula 12.5)}$$

where $Var\ 1_{high}$ is the higher variance among the two data set.
    The ratio follows F-distribution with (n_{high}-1, n_{low}-1) degrees of freedom.

Exercise 12.3
In a particular stock exchange, variance of share prices of firm A is 234, in a sample of 100 days, and variance of firm B is 210 in a sample of 150 days. Assess whether the variance is different.

Solution

$$\text{Var}_{high} = 234, n_{high} = 100$$

$$\text{Var}_{low} = 210, n_{low} = 150$$

$$F \text{ ratio} = \frac{234}{210} = 1.11$$

F value for degree of freedom 100–1, 150–1 for $\alpha = 0.05$ is 1.678.

Since the calculated F ratio is less than 1.678, both the variances are equal (homogenous samples).

12.5   Test for Inference: Two Small-Sized Samples (t-test)

If two small-sized samples are drawn, then it can be assessed whether they belong to same population by using t-test. t-statistics is given as

$$t = \frac{(\bar{x}_1 - \bar{x}_2) - 0}{\sqrt{s_t^2\left(\frac{1}{n_1} + \frac{1}{n_2}\right)}} \qquad \text{(Formula 12.6)}$$

where $s_t^2 = \frac{(n_1 s_1^2 + n_2 s_2^2)}{n_1 + n_2 - 2}$ and $df = n_1 + n_2 - 2$.

The t-statistics is valid only if the data is normally distributed, and both samples are homogenous, i.e. having equal variance (assessed by F-test described above).

Exercise 12.4
In a study, 15 patients in the test group were given a new iron supplement and 13 patients in control group were given a placebo. After 1 month, the mean haemoglobin of test group was 14.8 g/dl and SD 2.8, while that of the control group was 13.1 g/dl and SD 2.6. Determine if the difference between the two groups is statistically significant.

Solution

$$s_t^2 = \frac{(n_1 s_1^2 + n_2 s_2^2)}{n_1 + n_2 - 2}$$

$$s_t^2 = \frac{(15 * 2.8^2 + 13 * 2.6^2)}{15 + 13 - 2}$$

$$s_t^2 = 7.9$$

$$t = \frac{(\bar{x}_1 - \bar{x}_2) - 0}{\sqrt{s_t^2\left(\frac{1}{n_1} + \frac{1}{n_2}\right)}}$$

$$t = \frac{(14.8 - 13.1) - 0}{\sqrt{7.9\left(\frac{1}{15} + \frac{1}{13}\right)}}$$

$$t = \frac{1.7}{\sqrt{11.376}}$$

$$t = \frac{1.7}{3.37}$$

$$t = 0.50$$

Since calculated t is less than 2.056 (see appendix 2), the means of two samples are not significantly different from each other.

12.6   Test for Inference of Paired Samples (Paired t-test)

Paired samples are the ones where there is only one sample of objects/subjects with each having two observations. It is commonly used to study role of a factor when the observations are made before and after the application/effect of the said factor. Paired t-test is based on the assumption that the mean of resulting differences of paired observations should be zero (null hypothesis). The steps for this test are as follows:

1. Find the difference of each set of paired observation.
2. Calculate the mean of resulting differences.
3. Calculate the standard deviation (SD) of differences.
4. Calculate the standard error of (SE) of differences as follows:

$$SE = \frac{SD}{\sqrt{n}} \qquad \text{(Formula 12.7)}$$

where n is number of subjects/objects.

5. Determine t value by using.

$$t = \frac{\bar{x} - 0}{SE} \qquad \text{(Formula 12.8)}$$

6. Degree of freedom $df = n-1$.

Exercise 12.5
In a particular study, haemoglobin levels in 12 anaemic pregnant females were measured before and after 1 month of iron folic acid therapy. Determine whether there is significant difference in haemoglobin levels before and after therapy (Tables 12.1).

Table 12.1 Haemoglobin levels (g/dl) before and after therapy

S. no	Before therapy	After therapy
1.	7.2	9.2
2.	8.3	10.1
3.	9.0	9.5
4.	9.5	11.2
5.	8.9	11.4
6.	6.9	9.0
7.	7.4	8.3
8.	8.3	8.7
9.	8.7	8.8
10.	9.3	9.4
11.	8.9	9.9
12.	10.5	11.2

Solution (Table 12.2)

Mean of the differences in observations is 1.15.

SD of the differences in observations is 0.83.

$$SE = \frac{0.83}{\sqrt{12}} = 0.24$$

$$t = \frac{\bar{x} - 0}{SE}$$

$$t = \frac{1.15 - 0}{0.24} = 4.78$$

Since calculated t is more than 2.201, for $df=11$ (see appendix 2), the null hypothesis is rejected. This implies that after iron folic acid therapy, there is a significant difference in haemoglobin levels before and after the therapy.

Table 12.2 Calculation of difference of paired observations

S. no	Before therapy	After therapy	Difference
1.	7.2	9.2	2
2.	8.3	10.1	1.8
3.	9	9.5	0.5
4.	9.5	11.2	1.7
5.	8.9	11.4	2.5
6.	6.9	9	2.1
7.	7.4	8.3	0.9
8.	8.3	8.7	0.4
9.	8.7	8.8	0.1
10.	9.3	9.4	0.1
11.	8.9	9.9	1
12.	10.5	11.2	0.7

Test for Inference: Multiple Sample Comparisons

13

Abstract

Background
Concept of sum of squares
ANOVA one way
ANOVA two way
Duncan's multiple range method
Fisher's least significance difference test

Keywords

Concept of sum of squares · ANOVA one way · ANOVA two way · Duncan's multiple range · Fisher's least significance difference

13.1 Background

In practical situations of biomedical statistics, we commonly come across examples of comparisons of multiple samples to assess the significance of difference between them for various variables. Such comparisons can be done by one or more of the following methods:

1. ANOVA – one way and two way.
2. Duncan's multiple range method.
3. Fisher's LSD (least significant difference) method.
4. Bonferroni method.
5. Tukey's HSD (honest significant difference) method.

Before we go on to understand these, let us spend a little time in understanding the concept of sum of squares that is applied in these tests. The equations may seem heavy, but you do not need to memorize the equations themselves.

© Springer Nature Singapore Pte Ltd. 2019
S. K. Yadav et al., *Biomedical Statistics*,
https://doi.org/10.1007/978-981-32-9294-9_13

13.2   Concept of Sum of Squares

The sum of differences of each value (y_i) from their mean $(\bar{y})$ is known as "sum of squares" (SS). For single series $(with\ observation\ i = 1\ to\ n)$, sum of squares is given by.

$$SS = \Sigma(y_i - \bar{y}) \qquad \text{(Formula 13.1)}$$

For more than one series $(for\ series\ j = 1\ to\ k)$, sum of squares is given by.

$$(y_{ij} - \overline{Y}) = (y_{ij} - \overline{Y_j}) + (\overline{Y_j} - \overline{Y})$$

where $\bar{y}$ is grand mean and $\overline{y_j}$ is mean of series j Since

$$\Sigma_j\Sigma_i\left(y_{ij} - \overline{Y_j}\right)^2 = \Sigma_j\Sigma_i\left[\left(y_{ij} - \overline{Y_j}\right) + \left(\overline{Y_j} - \overline{Y}\right)\right]^2$$

$$= \Sigma_j\Sigma_i\left(y_{ij} - \overline{Y}\right)^2 + \Sigma_j\Sigma_i\left(\overline{Y_j} - \overline{Y}\right)^2 - 2\Sigma_j\Sigma_i\left(y_{ij} - \overline{Y_j}\right)\left(\overline{Y_j} - \overline{Y}\right)$$

$$2\Sigma_j\Sigma_i\left(y_{ij} - \overline{Y_j}\right)\left(\overline{Y_j} - \overline{Y}\right) = 2\Sigma_j\left(\overline{Y_j} - \overline{Y}\right)\Sigma_i\left(y_{ij} - \overline{Y_j}\right) = 0$$

and

$$\Sigma_j\Sigma_i\left(\overline{Y_j} - \overline{Y}\right)^2 = \Sigma_j n_j\left(\overline{Y_j} - \overline{Y}\right)^2$$

$$\Sigma_j\Sigma_i\left(y_{ij} - \overline{Y}\right)^2 = \Sigma_j\Sigma_i\left(y_{ij} - \overline{Y_j}\right)^2 + \Sigma n_j\left(\overline{Y_j} - \overline{Y}\right)^2$$

$$\text{Total } SS = \text{within } SS + \text{between } SS$$

$$\text{Abbreviated as TSS} = \text{WSS} + \text{BSS}$$

13.2.1 To Calculate ESS

$$\text{WSS or ESS or RSS} = \text{TSS} - \text{BSS}$$

$$\Sigma_j\Sigma_i\left(y_{ij} - \overline{Y_j}\right)^2 = \text{TSS} - \text{BSS}$$

where ESS is error sum of squares and RSS is residual sum of squares.

13.2.2 To Calculate TSS

Similar to the Formula 8.11, total sum of squares can be written as

$$\text{TSS} = \Sigma_j\Sigma_i\left(y_{ij} - \overline{Y}\right)^2 = \Sigma_j\Sigma_i\left(y_{ij}\right)^2 - \frac{\left(\Sigma_j\Sigma_i y_{ij}\right)^2}{N}$$

$$\text{TSS} = \frac{\Sigma_j\left(\Sigma_i y_{ij}\right)^2}{n_j} - CF$$

where $CF = \frac{\left(\Sigma_j\Sigma_i y_{ij}\right)^2}{N} = \frac{T^2}{N}$,

T is the grand total of all observations of various groups together and N is the total number of observations.

13.2.3 To Calculate BSS

$$\text{BSS} = \Sigma n_j\left(\overline{Y}_j - \overline{Y}\right)^2 = \frac{\Sigma_j\left(\Sigma_i y_{ij}\right)^2}{n_j} - CF$$

$$\text{BSS} = \Sigma_j n_j\left(\overline{Y}_j\right)^2 - CF$$

$$\text{BSS} = \Sigma_j\left[n_j\left(\frac{\Sigma_i y_{ij}}{n_j}\right)\left(\frac{\Sigma_i y_{ij}}{n_j}\right)\right] - CF$$

$$\text{BSS} = \Sigma_j\left[\frac{\left(\Sigma_i y_{ij}\right)\left(\Sigma_i y_{ij}\right)}{n_j}\right] - CF$$

$$\text{BSS} = \Sigma_j\left(\frac{T_j^2}{n_j}\right) - CF$$

T_j^2 is the square of total of the column "j"and, hence, can be written as T_{cj}^2.
Here sum of squares is calculated across the column. BSS can be written as BSS_c.

13.2.4 Special Scenario

If we have more than one series and each series have equal number of observations, then sum of squares across the rows is also possible.

$$\text{TSS} = \frac{\Sigma_j\left(\Sigma_i y_{ij}\right)^2}{n_j} - CF$$

$$BSS_c = \Sigma_j \left(\frac{T_{cj}^2}{n_j} \right) - CF$$

$$BSS_r = \Sigma_i \left(\frac{T_{ri}^2}{n_i} \right) - CF$$

$$ESS = TSS - BSS_c - BSS_r$$

13.3   ANOVA One Way

ANOVA stands for "Analysis of Variance" within and between groups of data set with multiple groups. Each term of the data can be decomposed into the effect of overall mean and the property of its group. The linear model is given by

$$y_{ij} = \mu + \alpha_j + e_{ij}$$

where y_{ij} is the ith value of the jth group, μ is overall mean, α_j is the effect due to being in j^{th} column and e_{ij} is the error term. All e_{ij} are independent and are normally distributed $N(0, \sigma)$.

Null Hypothesis
H_0: mean 1 = mean 2 for all pairs in a groups

Alternate Hypothesis
H_1: mean 1 $\neq$ mean 2 for one or more pairs in a group

Partitioning of variances is done.

$$Total\ SS = within\ SS + between\ SS$$

$\left[\frac{BSS}{(k-1)} \right]$ is between group variance and $\left[\frac{ESS}{N-k} \right]$ is within groups variance. These variances can be assessed using F-test as follows.

$$F = \frac{\left[\frac{BSS}{(k-1)} \right]}{\left[\frac{ESS}{N-k} \right]}$$

$$df = (k - 1), (N - k)$$

13.4   ANOVA Two Way

ANOVA two way is the analysis of variance within groups, across columns and across rows. The linear model is given by

$$y_{cr} = \mu + \alpha_c + \beta_r + e_{cr}$$

where μ is overall mean, α_c is the effect of being in cth column, β_r is the effect of being in rth row and e_{cr} is the error term.

Partitioning of variances is done.

$$\text{Total } SS = \text{within } SS + \text{between } SS_c + \text{between } SS_r$$

$\left[\frac{BSS_c}{(c-1)}\right]$ is between group variance across columns, $\left[\frac{BSS_r}{(r-1)}\right]$ is between group variance across rows and $\left[\frac{ESS}{(c-1)*(r-1)}\right]$ is within groups variance. These variances can be assessed using F-test as follows:

$$F_c = \frac{\left[\frac{BSS_c}{(c-1)}\right]}{\left[\frac{ESS}{(c-1)*(r-1)}\right]}$$

$$df = (c-1), (c-1)*(r-1)$$

and

$$F_r = \frac{\left[\frac{BSS_r}{(r-1)}\right]}{\left[\frac{ESS}{(c-1)*(r-1)}\right]}$$

$$df = (r-1), (c-1)*(r-1)$$

13.5   Duncan's Multiple Range Method

In this method the means are arranged in ascending order.

Where $i < j$, i and j being ranks of arranged means in ascending order,

$$W = q(r, N-g)\sqrt{\frac{WMS}{n}}$$

Where within mean sum of squares (WMS),

$$\text{WMS} = \frac{\text{ESS}}{N - k}$$

Here, n is the number of observations per treatment groups, r is the "distance" $(j - i)$ between the maximum and minimum means and q is from table of "Studentized range" [Ref. 1].

If $\overline{X} - \dot{X}_i > W$, then the difference is significant.

13.5.1 Another Approach

Let Sd is defined as

$$Sd^2 = 2\frac{\text{RMS}}{n}$$

then

$$W = q(r, N - g)Sd$$

where $\text{RMS} = \frac{\text{ESS}}{N-k}$

In both approaches W is calculated for each pair using different r values.

13.6   Fisher's Least Significance Difference Test

13.6.1 Approach 1

The test is a generalized version of the pair wise t-test, where least square difference (LSD) is defined as

$$\text{LSD} = \mid X_i - X_j \mid \sqrt{\text{WMS}\left(\frac{1}{n_i} + \frac{1}{n_j}\right)}$$

and is compared with the t-distribution with $(N-g)$ degrees of freedom, g being the number of groups.

13.6.2 Approach 2

$$\text{LSD} = (t\alpha, df \text{ of WSS})\left(\sqrt{\text{WMS}\left(\frac{1}{n_i} + \frac{1}{n_j}\right)}\right)$$

where df of WSS is $N-g$.

Reference

1. Kokoska S, Nevison C (1989) Least signficant studentized ranges for Duncan's test. In: Statistical tables and formulae. Springer texts in statistics. Springer, New York

Test for Inference: Categorical Data I   **14**

Abstract

Background
Tabulation of categorical data
Test for significance of individual proportion
Test for significance of difference between two proportions

Keywords

Categorical data · Test for proportions

14.1 Background

As we have read earlier, categorical data is qualitative data which cannot take a numerical value by itself. Before test for inference can be applied to such data, proper tabulation is important to apply the statistical test appropriately.

14.2 Tabulation of Categorical Data

Categorical data can be tabulated in multiple ways:

1. Number of cases possessing a particular attribute can be tabulated as shown in Table 14.1.

© Springer Nature Singapore Pte Ltd. 2019
S. K. Yadav et al., *Biomedical Statistics*,
https://doi.org/10.1007/978-981-32-9294-9_14

Table 14.1 Number of typhoid cases in two societies

Society	Typhoid cases	Normal subjects	Total
A	5	495	500
B	12	350	362

2. Another way of tabulation of the data shown above is by denoting the proportions of cases with a particular attribute, as shown in Table 14.2. Proportions of cases with a particular attributes are considered to be better than absolute numbers since the proportion gives a better and clearer picture of the difference of the same attribute between two populations.

3. In certain circumstances, two attributes are studied for a particular group of cases (sample). In such a scenario, a special type of frequency table with cross-tabulation of categorical variables is used, which is known as "contingency table". An example is shown in Table 14.3. Such tabulation is useful to assess the association between the two attributes in a given population or sample.

Table 14.2 Proportions of typhoid cases in two societies

Society	Proportion	Total (n)
A	5/500	500
B	12/362	362

Both these tabulations can be used to assess the significance of difference in typhoid cases between society A and B

Table 14.3 Study of typhoid cases and use of water purifier in society A

	Typhoid cases	Normal subjects	Total
Using water purifier	1	299	300
Not using water purifier	4	196	200
Total	5	495	500

14.3   Test for Significance of Individual Proportion

Proportion is the number of cases with a particular type of attribute divided by total number of cases in the sample or population. As previously described, the statistics. $z = \frac{[t-E(t)]}{\sqrt{Var(t)}}$ follows normal distribution (0,1).

Putting $E(t) = nP$ and $Var(t) = np(1 - p)$ (see Chap. 27), where P is the proportion of cases in population with favourable attribute, we get.

$$z = \frac{[X - nP]}{\sqrt{[np(1 - p)]}}$$

where X is number of cases in sample with the favourable attribute, hence p is the proportion of cases in sample with favourable attribute.

Let $p = \frac{X}{n}$.

$$z = \frac{[p - P]}{\sqrt{\frac{[p(1-p)]}{n}}}$$

$$z = \frac{[p - P]}{\sqrt{\left[\frac{pq}{n}\right]}} \qquad \text{(Formula 14.1)}$$

where $q = 1 - p$.

Exercise 14.1
From various previous studies, it is known that diarrhoea is seen in 45% cases of cholera infection. In a recent study, 30% of 60 cholera patients gave a history of diarrhoea. Assess whether the findings of the recent study are consistent with the previous information.

Solution
Using Formula 14.1

$$z = \frac{[p - P]}{\sqrt{\left[\frac{PQ}{n}\right]}}$$

$$z = \frac{[30 - 45]}{\sqrt{\left[\frac{30*70}{60}\right]}}$$

$$z = -2.54$$

Since the calculated value of z is < -1.96, findings of present study are significantly different from previous studies.

14.4   Test for Significance of Difference Between Two Proportions

The same attribute may need to be compared between two populations or two samples. This is done using the test for significance of difference between two proportions.

Considering that the statistics

$$z = \frac{[t - E(t)]}{\sqrt{\text{Var}(t)}}$$

follows normal distribution $(0,1)$,
putting $t = p_1 - p_2$, $E(t) = 0$

$$\text{Var}(t) = \frac{p_1(1 - p_1)}{n_1} + \frac{p_2(1 - p_2)}{n_2},$$

it can be shown that

$$z = \frac{\{p_1 - p_2\}}{\sqrt{\left\{\frac{p_1(1-p_1)}{n_1} + \frac{p_2(1-p_2)}{n_2}\right\}}}$$

$$z = \frac{\{p_1 - p_2\}}{\sqrt{\left\{\frac{p_1 q_1}{n_1} + \frac{p_2 q_2}{n_2}\right\}}} \qquad \text{(Formula 14.2)}$$

where $q_1 = 1 - p_1$ and $q_2 = 1 - p_2$.

Exercise 14.2
In group A, myopia was found in 30 out of 60 children, and in group B, myopia was found in 90 out of 450 children. Find if the difference observed between the two groups is by chance.

Solution
Using Formula 14.2

$$z = \frac{\{p_1 - p_2\}}{\sqrt{\left\{\frac{p_1 q_1}{n_1} + \frac{p_2 q_2}{n_2}\right\}}}$$

$$n_1 = 60, n_2 = 450$$

$$p_1 = \frac{30}{60} * 100 = 50, p_2 = \frac{90}{450} * 100 = 20$$

$$q_1 = 50, q_2 = 80.$$

$$z = \frac{\{50 - 20\}}{\sqrt{\left\{\frac{50 \ast 50}{60} + \frac{20 \ast 80}{450}\right\}}}$$

$$z = 4.46$$

The calculated z is more than 1.96; hence, findings of the two groups are significantly different at $\alpha = 0.05$ and are not by chance.

Test for Inference: Categorical Data II

15

Abstract

Chi-square Test

Keywords

Chi-square test · Observed frequency · Expected frequency · Goodness of fit · Yates correction

15.1 Chi-Square Test

15.1.1 R x C Table

The most common statistical test for inference applied to categorical data is the chi-square test. For this test, the expected frequencies for the parameter to be evaluated are calculated by the following equation:

$$\text{Expected Frequency of a cell } (ij) = \frac{\text{Total of } i^{th} \text{ row} * \text{Total of } j^{th} \text{ column}}{\text{Total no.of observations.}}$$

The calculated expected frequencies are tabulated along with the observed frequencies, as explained below (Tables 15.1 and 15.2):

where R_i represents total of the ith row and C_j represents total of the jth column. The statistics

$$\chi^2 = \Sigma_i \Sigma_j \left[\frac{(O_{ij} - E_{ij})^2}{E_{ij}} \right] \qquad \text{(Formula 15.1)}$$

follows chi-square distribution with $df = (r - 1)(c - 1)$.

Degree of freedom is equal to (no. or rows – 1) (no. of columns – 1)

© Springer Nature Singapore Pte Ltd. 2019

S. K. Yadav et al., *Biomedical Statistics*,

https://doi.org/10.1007/978-981-32-9294-9_15

Table 15.1 Representation of observed frequencies

	Column 1	Column 2	Column 3	Total (R_i)
Row 1	O_{11}	O_{12}	O_{13}	R_1
Row 2	O_{21}	O_{22}	O_{23}	R_2
Row 3	O_{31}	O_{32}	O_{33}	R_3
Total	C_1	C_2	C_3	N

Table 15.2 Representation of expected frequencies

	Column 1	Column 2	Column 3	Total (R_i)
Row 1	E_{11}	E_{12}	E_{13}	R_1
Row 2	E_{21}	E_{22}	E_{23}	R_2
Row 3	E_{31}	E_{32}	E_{33}	R_3
Total	C_1	C_2	C_3	N

15.1.1.1 Assumptions for Chi-Square Test

Before applying the chi-square test, it has to ensure that the lowest expected frequency in any cell is not less than 5. If any of the value is less than 5, Fisher's exact test should be applied (by reducing the table to 2x2).

15.1.2 1 x C Tables (No. of Rows = 1, No. of Columns = >1)

Such a table is used in the situation when the observed frequencies of cases belonging to a single attribute are to be analysed against the already known frequencies. As the name suggests, there is only one row with multiple columns.

In this situation, the chi-square test is also known as test for "goodness of fit". It has degree of freedom $(c-1)$, since there is only one row. The formula for the chi-square statistics in this situation is as follows:

$$\chi^2 = \Sigma_j \left[\frac{(O_j - E_j)^2}{E_j} \right]$$

(Formula 15.2)

This test is frequently used to assess the results of data modelling into a particular statistical distribution, as illustrated in the exercise 15.1.

Exercise 15.1

Assess whether the observation of the present study fits the distribution as reported by a previous large study in the following data (Table 15.3).

Table 15.3 Data of distribution of blood groups in exercise 15.1

Blood group	A	B	AB	O
Present study (O)	16	26	10	40
Previous study (E)	80	96	46	110

Solution

$$\chi^2 = \frac{(O-E)^2}{E}$$

$$\chi^2 = \frac{(16-80)^2}{80} + \frac{(26-96)^2}{96} + \frac{(10-46)^2}{46} + \frac{(40-110)^2}{110}$$

$$\chi^2 = \frac{(64)^2}{80} + \frac{(70)^2}{96} + \frac{(36)^2}{46} + \frac{(70)^2}{110}$$

$$\chi^2 = \frac{4096}{80} + \frac{4900}{96} + \frac{1296}{46} + \frac{4900}{110}$$

$$\chi^2 = 51.2 + 51.04 + 28.17 + 44.54$$

$$\chi^2 = 174.95$$

At $df = 3$, $\chi^2 = 7.82$ for $\alpha = 0.05$ (see appendix 3)
7.82 < <174.95
Hence, the distribution of present study does not fit that of previous study.

15.1.3 2x2 Table

This is a special type of R X C table where the number of rows as well as columns is two. This cross-tabulated table is also known as "contingency table". It is a special case where Yates correction is advised in chi-square test. It is a correction for continuity.

$$\chi^2 = \Sigma_i \Sigma_j \left[\frac{(O_{ij} - E_{ij} - 0.5)^2}{E_{ij}} \right] \qquad \text{(Formula 15.3)}$$

An easy formula for chi-square for 2x2 contingency table is

$$\chi^2 = \frac{\left\{ (|ad - bc| - \frac{N}{2})^2 * N \right\}}{\{R_1 * R_2 * C_1 * C_2\}} \qquad \text{(Formula 15.4)}$$

where observed frequencies are labelled as shown in Table 15.4.

Table 15.4 Representation of observed frequencies to be used in formula 15.4

	Column 1	Column 2	Total
Row 1	a	b	R_1
Row 2	c	d	R_2
Total	C_1	C_2	N

Table 15.5 Contingency table

	Results		
Groups	Dryness present	Dryness absent	Total
Control (not using mobile)	12	23	35
Test (using mobile)	8	57	65
Total	20	80	100

Exercise 15.1

Find the association between dryness in eyes and prolonged mobile usage from the data given below (Table 15.5).

Solution

Using formula 15.4 for chi-square

$$\chi^2 = \frac{\left\{ (|ad - bc| - \frac{N}{2})^2 \times N \right\}}{\{R_1 * R_2 * C_1 * C_2\}}$$

$$\chi^2 = \frac{\left\{ (|12 * 57 - 23 * 8| - \frac{100}{2})^2 * 100 \right\}}{\{35 * 65 * 20 * 80\}}$$

$$\chi^2 = 5.56$$

At $df = 1$, $\chi^2 = 3.84$ for $\alpha = 0.05$ (see appendix 3)

Since the calculated χ^2 of 5.56 is higher than 3.84, it can be inferred that the prolonged use of mobile phones is significantly associated with dryness of eyes.

Test for Inference: Correlation and Regression

16

Abstract

Background
Pearson's correlation coefficient
Regression analysis

Keywords

Pearson's correlation · Least square regression

16.1 Background

In real-life situations, two variables may seem to be associated with each other. However, the association has to be assessed statistically since what seems may actually not be true. An association between variables with numerical data is commonly known as correlation. The correlation may be positive, i.e. when one variable rises and the other variable also increases, or negative, implying that the other variable may decrease; or the two variables may not be correlated at all when change in one variable does not have a corresponding effect on the other variable.

We have to remember that having a correlation between the two variables does not imply a cause-and-effect relationship between them. The presence of correlation simply implies that the two are related in some manner. For this correlation to be meaningful, the two variables must be logically related to each other. Apparent, spurious correlation may be present between two unrelated variables. An example of spurious correlation is a supposition that the forest cover on earth is gradually depleting as a person is growing older. However, there is no logical link between somebody growing older and depletion of the forest cover. Traditionally, correlation ranges from -1 (perfect negative) through 0 (no correlation) to $+1$ (perfect positive).

In order to assess the nature of correlation between two variables, the data is plotted on x-axis and y-axis and is represented by scattered dots (Fig. 16.1).

© Springer Nature Singapore Pte Ltd. 2019
S. K. Yadav et al., *Biomedical Statistics*,
https://doi.org/10.1007/978-981-32-9294-9_16

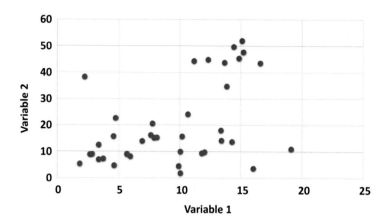

Fig. 16.1 Scatter plot

There are statistical tests to identify correlation between two variables as well as to quantify the correlation itself.

16.2 Pearson's Correlation Coefficient

Pearson's correlation coefficients measure relationship between two variables. It is calculated by using product moment of two variables, i.e. $\Sigma(x_i - \bar{x})(y_i - \bar{y})$. Correlation coefficient is denoted by r, and it can vary from -1 to $+1$, where -1 indicates perfect inverse correlation and $+1$ indicates perfect direct correlation:

$$r = \frac{\Sigma(x_i - \bar{x})(y_i - \bar{y})}{\sqrt{\Sigma(x_i - \bar{x})^2 \Sigma(y_i - \bar{y})^{2\cdot}}}$$

$$r = \frac{\Sigma xy - \frac{\Sigma x \Sigma y}{n}}{\sqrt{\left(\Sigma x^2 - \frac{(\Sigma x)^2}{n}\right)\left(\Sigma y^2 - \frac{(\Sigma y)^2}{n}\right)}}$$

$$r = \frac{n\Sigma xy - \Sigma x \Sigma y}{\sqrt{\left(n\Sigma x^2 - (\Sigma x)^2\right)\left(n\Sigma y^2 - (\Sigma y)^2\right)}} \qquad \text{(Formula 16.1)}$$

16.2.1 Test for the Significance of *r*

16.2.1.1 Small Sample (t-test)

$$t = \frac{[r - 0]}{[SE(r)]} = \frac{r}{\sqrt{\left[\frac{(1-r^2)}{n-2}\right]}}$$

Here "t" follows Student's t-distribution with *n-2* degrees of freedom.

16.2.1.2 Large Sample (z-test)

$$z = \frac{\left[\tan h^{-1} r - \tan h^{-1} \rho\right]}{\sqrt{\left[\frac{1}{n-3}\right]}}$$

Here "z" follows normal distribution (0,1)

where *r* is the calculated correlation coefficient, ρ is the known population correlation coefficient and *n* is the number of paired observations.

16.3   Regression Analysis

Regression analysis is the quantitative study of change in a dependent variable due to change in one or more independent variable(s). The two variables may be related to each other either linearly or in a nonlinear manner.

When two variables, *x* and *y*, are related linearly, their relationship can mathematically be expressed in terms of equation of a straight line. For nonlinear relation, equation of a polynomial is utilized. Equation of a straight line can be written as follows:

$$y = a + bx$$

where *y* is related to *x* by constant *a* and slope *b*. Parameters *a* and *b* can be estimated by either least square regression or orthogonal regression methods.

By using least square regression

$$b = \frac{[\Sigma xy - - n\bar{x}\bar{y}]}{[\Sigma x^2 - - n\bar{x}^2]}$$

$$b = \frac{\left[\Sigma xy - \frac{(\Sigma x \, \Sigma y)}{n}\right]}{\left(\Sigma x^2 - \frac{(\Sigma x)^2}{n}\right)}$$

$$b = \frac{n\Sigma xy - \Sigma x\Sigma y}{n\Sigma x^2 - \Sigma x\Sigma x} \qquad \text{(Formula 16.2)}$$

$$a = \bar{y} - b\bar{x}$$

$$a = \frac{\Sigma y - b\Sigma x}{n}$$

If y is related to x, then x can also be estimated from y as follows:

$$x = a' + b'y$$

$$b' = \frac{[\Sigma xy - -n\overline{xy}]}{[\Sigma y^2 - -n\bar{y}^2]}$$

$$b' = \frac{\left[\Sigma xy - \frac{(\Sigma x\Sigma y)}{n}\right]}{\left(\Sigma y^2 - \frac{(\Sigma y)^2}{n}\right)}$$

$$a' = \bar{x} - -b\bar{y}$$

It can be shown that $bb' = r^2$.

16.3.1 Significance of Model

Standard error of regression is given as

$$\sigma_{est} = \sqrt{\frac{\Sigma(y_i - \hat{y}_i)^2}{n - 2}}$$

Standard error of intercept is

$$SE_a = \frac{s}{\sqrt{n}}\sqrt{1 + \frac{\bar{x}^2}{\sigma_x^2}}$$

Standard error of β is

$$SE_\beta = \frac{s}{\sqrt{n}}\sqrt{\frac{1}{\sigma_x^2}}$$

To assess the significance of coefficient, t-test is given as

$$t = \frac{\beta}{SE_\beta}$$

For degree of freedom $n-2$

To assess the significance of model, F-test is given as

$$F = \left[\frac{\Sigma(\hat{y}_i - \bar{y})^2}{p - 1}\right] * \left[\frac{(n - p)}{\Sigma(y_i - \hat{y}_i)^2}\right]$$

For degree of freedom $(p-1, n-p)$ for univariate model.

Where p are number of parameters (two, α and β, for univariate model) and n is number of observations.

Coefficient of determination is given as

$$R^2 = 1 - \frac{\sigma^2_{est}}{\sigma^2_y}$$

$$AdjR^2 = 1 - \left[\frac{(n - 1)}{(n - 2)} * (1 - R^2)\right]$$

16.3.2 Data Presentation

When one variable is regressed over the other, an equation for a graph is obtained. Regression may be linear resulting in a straight line or nonlinear resulting in a polygon. The entire data may or may not be shown as scatter plot along with regression line (Fig. 16.2).

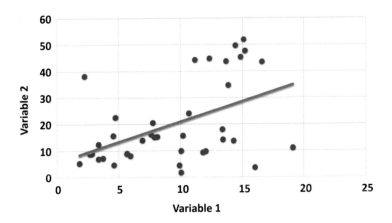

Fig. 16.2 Scatter plot with regression line

Table 16.1 Haemoglobin values with their corresponding mean corpuscular volume

Haemoglobin (x)	MCV (y)
8.4	70.2
9.3	71.3
7.6	68.4
12.9	83.1
15.3	90.2
10.1	72.4
8.1	69.3
7.9	68.9

Table 16.2 Preparation for calculation of product moment

Haemoglobin (x)	MCV (y)	xy	x^2	y^2
8.4	70.2	589.68	70.56	4928.04
9.3	71.3	663.09	86.49	5083.69
7.6	68.4	519.84	57.76	4678.56
12.9	83.1	1071.99	166.41	6905.61
15.3	90.2	1380.06	234.09	8136.04
10.1	72.4	731.24	102.01	5241.76
8.1	69.3	561.33	65.61	4802.49
7.9	68.9	544.31	62.41	4747.21
Total 79.6	593.8	6061.54	845.34	44523.4

Exercise 16.1

In a particular study, red cell indices were noted. The data for haemoglobin and red cell mean corpuscular volume (MCV) is given in the table below (Table 16.1). Determine the Pearson correlation coefficient between haemoglobin and mean corpuscular volume and find its significance.

Solution (Table 16.2)

$$r = \frac{\Sigma xy - \frac{\Sigma x \Sigma y}{n}}{\sqrt{\left(\Sigma x^2 - \frac{(\Sigma x)^2}{n}\right)\left(\Sigma y^2 - \frac{(\Sigma y)^2}{n}\right)}}$$

$$r = \frac{6061.54 - \frac{79.6*593.8}{8}}{\sqrt{\left(845.34 - \frac{(79.6)^2}{8}\right)\left(\Sigma 44523.4 - \frac{(593.8)^2}{8}\right)}}$$

$$r = \frac{153.23}{\sqrt{53.32 * 448.59}}$$

$$r = \frac{153.23}{\sqrt{23919.09}}$$

$$r = \frac{153.23}{154.66}$$

$$r = 0.99$$

For Significance

$$t = \frac{r}{\sqrt{\left[\frac{(1-r^2)}{n-2}\right]}}$$

$$t = \frac{0.99}{\sqrt{\left[\frac{(1-0.99^2)}{8-2}\right]}}$$

$$t = \frac{0.99}{\sqrt{\left[\frac{(1-0.99^2)}{8-2}\right]}}$$

$$t = \frac{0.99}{\sqrt{\left[\frac{0.02}{6}\right]}}$$

$$t = \frac{0.99}{0.054}$$

$$t = 18.33$$

Since calculated t is more than 2.447 (value of t *see appendix 2 for df = 6 at α = 0.05*), the correlation coefficient is statistically significant.

Exercise 16.2
For data given in Exercise 16.1, do a regression analysis.
   Solution

$$y = a + bx$$

$$b = \frac{\left[\Sigma xy - \frac{(\Sigma x \, \Sigma y)}{n}\right]}{\left(\Sigma x^2 - \frac{(\Sigma x)^2}{n}\right)}$$

$$b = \frac{153.23}{53.32} = 2.874$$

$$a = \bar{y} - b\bar{x}$$

$$a = 74.225 - 2.874 * 9.95$$

$$a = 45.6287$$

$$y = 45.6287 + 2.874x$$

$$MCV = 45.6287 + 2.874Hb$$

Nonparametric Tests

17

Abstract

Background
Mann-Whitney U test
Wilcoxon signed rank sum test
Kruskal-Wallis H test
Spearman Rank Correlation

Keywords

Mann-Whitney U · Wilcoxon signed rank sum · Kruskal-Wallis H · Spearman
rank correlation

17.1   Background

When the data does not have a normal distribution, the tests prescribed so far in this
book regarding significance of mean or difference between one or more group means
cannot be used, since most of them assume data to be normally distributed. For such
a data, nonparametric tests are used which are described in this chapter. These tests
are based on rank conversion of data. The frequently used nonparametric test in
biomedical statistics is tabulated in Table 17.1.

17.1.1 Mathematical Basis of Rank Tests

These tests are based on ranking of the arranged observations. Some general
calculations of ranks of n consecutive numbers are given as follows:

Table 17.1 Nonparametric tests

Nonparametric tests
Mann-Whitney U test
Kruskal-Wallis H test
Wilcoxon signed rank sum test
Spearman rank correlation

$$\text{Sum of } n \text{ consecutive numbers} = \frac{n(n+1)}{2}$$

$$\text{Mean of } n \text{ consecutive numbers} = \left(\frac{n(n+1)}{2}\right) * \frac{1}{n} = \frac{n+1}{2}$$

$$\text{variance of } n \text{ consecutive numbers} = \frac{n^2 - 1}{12}$$

17.2 Mann-Whitney U Test

Mann-Whitney U test is used in place of unpaired t-test for data which is not normally distributed. Null hypothesis for this test is that the two samples belong to the same population. In Mann-Whitney U test, both the samples (sample 1 and sample 2) are combined to get a total of $n_1 + n_2$ observations. These observations are then arranged from lowest to highest and are ranked. For ties, average rank is assigned. These ranks are then separated into group 1 and group 2 corresponding to sample 1 and sample 2.

Sum of all the combined ranks will be equal to

$$\frac{N(N+1)}{2}$$

where $N = n_1 + n_2$

U statistics is defined as

$$U_1 = n_1 n_2 + \left[\frac{n_1(n_1 + 1)}{2}\right] - T_1 \qquad \text{(Formula 17.1)}$$

where T_1 is the observed sum of ranks of sample 1.

The significance of U statistics can be assessed from the tabulated distribution of U. For larger (more than 20) samples normal approximation can be used to assess significance using z-test.

$$z = \frac{U - \text{Exp}(U)}{SD(U)}$$

$$\text{Exp}(U) = \frac{n_1 n_2}{2}$$

$$SD(U) = \sqrt{\frac{n_1 n_2 (n_1 + n_2 + 1)}{12}}$$

The significance of z can be looked up in the table of standard normal distribution.

Exercise 17.1

Assess whether haemoglobin levels given in Table 17.2 of group A and B are significantly different.

Table 17.2 Haemoglobin levels (g/dl) of group A and B

Group A	Group B
8.5	6
9	6.5
9.2	6.8
9.5	7.4
11	7.6
11.5	7.7
11.9	8.1
12	8.2
12.4	8.5
12.6	9

Solution

After pooling the data and arranging in order, ranks of two groups are shown in Table 17.3.

Table 17.3 Ranks of group A and B obtained after pooling of data

Group A	Rank	Group B	Rank
8.5	9	6	1
9	12	6.5	2
9.2	13	6.8	3
9.5	14	7.4	4
11	15	7.6	5
11.5	16	7.7	6
11.9	17	8.1	7
12	18	8.2	8
12.4	19	8.5	10
12.6	20	9	11
Rank sum	153		57

$$n_1 = 10, n_2 = 10$$

$$T_1 = 153$$

$$U_1 = n_1 n_2 + \left[\frac{n_1(n_1 + 1)}{2}\right] - T_1$$

$$U_1 = 10 * 10 + \left[\frac{10(10 + 1)}{2}\right] - 153$$

$$U_1 = 100 + 55 - 153$$

$$U_1 = 2$$

$$\text{Exp}(U) = \frac{n_1 n_2}{2}$$

$$\text{Exp}(U) = \frac{10 * 10}{2} = 50$$

$$SD(U) = \sqrt{\frac{n_1 n_2 (n_1 + n_2 + 1)}{12}}$$

$$SD(U) = \sqrt{\frac{10 * 10(10 + 10 + 1)}{12}}$$

$$SD(U) = \sqrt{175} = 13.22$$

$$z = \frac{U - Exp\,(U)}{SD(U)}$$

$$z = \frac{2 - 50}{13.22} = -3.63$$

Since the calculated z is less than -1.96, the difference between group A and B is statistically significant.

17.3 Wilcoxon Signed Rank Sum Test

It is a nonparametric alternative for one sample t-test or paired t-test. Here W statistics is calculated as follows.

Scenario 1

One sample test: **for this we require a hypothesized median (or population median, M)**
1. Calculate the difference of each observation from the hypothesized median.

$$d_i = x_i - M$$

2. Assign rank to all differences ignoring the sign.
3. Apply the sign of d_i to its corresponding rank.
4. Calculate W^+ = sum of positive ranks, W^- =sum of negative ranks. Here.

$$W^+ + W^- = \frac{n(n+1)}{2}$$

Scenario 2

Paired Sample Test

1. Calculate the difference of paired observation.

$$d_i = x_i - y_i$$

2. Assign rank to all differences ignoring the sign.
3. Apply the sign of d_i to its corresponding rank.
4. Calculate W^+ = sum of positive ranks, W^- = sum of negative ranks. Here.

$$W^+ + W^- = \frac{n(n+1)}{2}$$

$$W = \min(W^+, W^-) \qquad \text{(Formula 17.2)}$$

The significance of W statistics can be assessed from the tabulated distribution of W, or for larger (more than 20) samples, normal approximation can be used to assess significance using z-test.

$$z = \frac{W - \text{Exp}(W)}{SD(W)}$$

$$\text{Exp}(W) = \frac{n(n+1)}{4}$$

$$SD(W) = \sqrt{\frac{n(n+1)(2n+1)}{24}}$$

The significance of z can be looked up in the table of standard normal distribution (See Appendix 1).

Exercise 17.2
Assess whether the median of group A (Table 17.2) is significantly different from population median of 11.25.

Solution
The differences from population median are shown in Table 17.4.

$$W^+ = \text{sum of positive ranks} = 20, \; W^- = \text{sum of negative ranks} = 35$$

$$\text{Exp}(W) = \frac{10(10+1)}{4} = 27.5$$

$$SD(W) = \sqrt{\frac{n(n+1)(2n+1)}{24}}$$

$$SD(W) = \sqrt{\frac{10(10+1)(2*10+1)}{24}}$$

$$SD\,(W) = 9.81$$

$$z^+ = \frac{W - \text{Exp}\,(W)}{SD(W)}$$

$$z^+ = \frac{20 - 27.5}{9.81}$$

$$z^+ = 0.764$$

Hence, the median of the group A is not significantly different from population median.

Table 17.4 Differences from population median and ranks

Group A	Population median	Difference	Absolute difference	Rank
8.5	11.25	−2.75	2.75	10
9	11.25	−2.25	2.25	9
9.2	11.25	−2.05	2.05	8
9.5	11.25	−1.75	1.75	7
11	11.25	−0.25	0.25	1
11.5	11.25	0.25	0.25	2
11.9	11.25	0.65	0.65	3
12	11.25	0.75	0.75	4
12.4	11.25	1.15	1.15	5
12.6	11.25	1.35	1.35	6

Exercise 17.3

Assess whether haemoglobin levels before and after (Table 17.5) hematinic treatment are significantly different.

Solution

Table differences and corresponding rank are tabulated in Table 17.6.

W^+ = sum of positive ranks = 4, W^- = sum of negative ranks = 40.

$$\mathrm{Exp}(W) = \frac{9(9+1)}{4} = 22.5$$

$$SD(W) = \sqrt{\frac{n(n+1)(2n+1)}{24}}$$

Table 17.5 Haemoglobin levels before and after hematinic treatment

Before	After
9	10.6
10	11.5
11	12.3
13	14
11.5	12.4
12.3	13.1
12.2	12.5
11.9	11.4
12.6	12

Table 17.6 Difference between before and after haemoglobin levels with their ranks

Before	After	Difference	Absolute difference	Rank
9	10.6	−1.6	1.6	9
10	11.5	−1.5	1.5	8
11	12.3	−1.3	1.3	7
13	14	−1	1	6
11.5	12.4	−0.9	0.9	5
12.3	13.1	−0.8	0.8	4
12.2	12.5	−0.3	0.3	1
11.9	11.4	0.5	0.5	2
12.6	12	0.6	0.6	3

$$SD(W) = \sqrt{\frac{9(9+1)(2*9+1)}{24}}$$

$$SD\ (W) = 8.44$$

$$z^+ = \frac{W - \mathrm{Exp}\ (W)}{SD(W)}$$

$$z^+ = \frac{5 - 22.5}{8.44}$$

$$z^+ = -2.073$$

Hence, there is significant difference in haemoglobin levels before and after treatment.

17.4   Kruskal-Wallis H Test

Kruskal-Wallis H test is used instead of ANOVA one way for data which is not normally distributed. In Kruskal-Wallis H test, samples of all the k groups are combined to get total observations. These observations are then arranged from lowest to highest and are ranked. For ties, average rank is assigned. These ranks are then separated into various groups.
   Sum of all the combined ranks will be equal to

$$\frac{N(N+1)}{2}$$

where

$$N = \Sigma n_i$$

$$H = (N-1)\frac{\Sigma n_i \left(\overline{R}_i - \overline{R}\right)^2}{\Sigma_i \Sigma_j \left(R_{ij} - \overline{R}\right)^2}$$

Since

$$\Sigma_i \Sigma_j \left(R_{ij} - \overline{R}\right)^2 = \frac{(n-1)(n)(n+1)}{12}$$

Hence,

$$H = \frac{12\Sigma n_i (\overline{R_i} - \overline{R})^2}{N(N+1)} \qquad \text{(Formula 17.3)}$$

where

$$\overline{R} = \frac{(N+1)}{2}$$

where R_i is the observed sum of ranks of group i.

The significance of H statistics can be assessed from the Chi-square distribution with $k - 1$ degree of freedom.

Exercise 17.4

Assess whether the haemoglobin levels between group A, B and C (Table 17.7) are significantly different.

Solution

Ranks of various groups obtained after poling of data are tabulated in Table 17.8.

$$H = \frac{12\Sigma n_i (\overline{R_i} - \overline{R})^2}{N(N+1)}$$

$$\Sigma n_i (\overline{R_i} - \overline{R})^2 = 429$$

$$H = \frac{12 * 429}{18(18+1)}$$

$$H = 15.05$$

Chi-square for $df = 2$, $\alpha = 0.05$ is 5.99. Since calculated $H > 5.99$, the difference in haemoglobin levels between groups is significant.

Table 17.7 Haemoglobin levels of group A, B and C

Group A	Group B	Group C
6.5	12	4.5
6.7	12.5	5.5
7.2	11.9	5
7.8	12.6	4
8.1	12.9	3.9
11	13	
	11.6	

Table 17.8 Ranks of group A, B and C

Group A	Rank A	Group B	Rank B	Group C	Rank C
6.5	6	12	14	4.5	3
6.7	7	12.5	15	5.5	5
7.2	8	11.9	13	5	4
7.8	9	12.6	16	4	2
8.1	10	12.9	17	3.9	1
11	11	13	18		
		11.6	12		
n_i	6		7		5
Sum	51		105		15
Mean $\overline{R_i}$	8.5		15		3

17.5 Spearman Rank Correlation

This is a nonparametric alternative of Pearson's correlation coefficient. Spearman rank correlation gives value ranging from -1 to $+1$, where -1 denotes a perfect negative correlation and $+1$ denotes a perfect positive correlation, while 0 denotes no correlation between the ranks.

For this, observations from both the series are ranked independently. Then difference (d_i) of paired ranks is calculated:

$$d_i = x_i - y_i$$

Spearman rank correlation is given by

$$\rho = 1 - \frac{6\Sigma d_i^2}{n(n^2 - 1)} \qquad \text{(Formula 17.4)}$$

Statistical significance of ρ can be assessed similar to that for Pearson's r (see Chap. 16).

Basis of Spearman correlation formula

$$d_i = x_i - y_i$$

$$d_i = (x_i - \bar{x}) - (y_i - \bar{y})$$

$$\Sigma d_i^2 = \Sigma\{(x_i - \bar{x}) - (y_i - \bar{y})\}^2$$

$$\frac{1}{n}\Sigma d_i^2 = \sigma_x^2 + \sigma_y^2 - 2\,Cov(x, y)$$

where

$$Cov\,(x,\,y) = \frac{\Sigma(x_i - \bar{x})(y_i - \bar{y})}{n}$$

Since

$$r = \frac{Cov(x, y)}{\sigma_x \sigma_y}$$

$$Cov(x, y) = r\sigma_x \sigma_y$$

$$\frac{1}{n}\Sigma d_i^2 = \sigma_x^2 + \sigma_y^2 - 2r\sigma_x \sigma_y$$

In our case

$$\sigma_x^2 = \sigma_y^2$$

$$r = 1 - \frac{\Sigma d_i^2}{2n\sigma_x^2}$$

Since

$$\sigma_x^2 = \frac{n^2 - 1}{12}$$

$$r = 1 - \frac{6\Sigma d_i^2}{n(n^2 - 1)}$$

Exercise 17.5
Assess the magnitude of correlation between group A and B of data given in Table 17.9.

Group A	Group B
5	11
6	11.5
7	11.9
8	12
7.5	12.4
10	13
9	13.6
9.5	14

Table 17.9 Values of group A and B

Solution

Group A and B are ranked independently, and their difference is shown in Table 17.10.

$$\rho = 1 - \frac{6\Sigma d_i^2}{n(n^2 - 1)}$$

$$\rho = 1 - \frac{6 * 8}{8(8^2 - 1)}$$

$$\rho = 1 - \frac{48}{504}$$

$$\rho = 1 - 0.095$$

$$\rho = 0.904$$

Magnitude of correlation between group A and group B is 0.904.

Table 17.10 Ranks of group A and B

Group A	Rank A	Group B	Rank B	Difference	Difference^2 (d_i^2)
5	1	11	1	0	0
6	2	11.5	2	0	0
7	3	11.9	3	0	0
8	5	12	4	1	1
7.5	4	12.4	5	−1	1
10	8	13	6	2	4
9	6	13.6	7	−1	1
9.5	7	14	8	-1	1
Total Σd_i^2					8

Sample Size Estimation

18

Abstract

Background
Sample size for assessing significance of single mean
Sample size for assessing significance of single proportion
Sample size for assessing significance of two means
Sample size for assessing significance of two proportion
Sample size for correlation coefficient
Sample size for sensitivity and specificity
Sample size for univariate logistic regression
Relative risk
Sample size to assess significance to two incidences in a cohort study
Sample size for case control and cohort studies
Sample size for unequal samples
Sample size for finite population

Keywords
Sample size calculation · Sample size formula

18.1 Background

We have learnt earlier that a sample drawn from a population is meant to represent the population itself and the results derived from this sample are usually extrapolated to the population. Hence, it is quite imperative to be prudent in calculating the sample size before a study is initiated. Sample size also affects the type II error and in turn the power of study. The method of calculation of sample size would differ as per the parameter under evaluation and the existing literature on that particular parameter which is used for this calculation for the present study.

© Springer Nature Singapore Pte Ltd. 2019
S. K. Yadav et al., *Biomedical Statistics*,
https://doi.org/10.1007/978-981-32-9294-9_18

18.2   Sample Size for Assessing Significance of Single Mean

If the investigator has available with him/her results of a previous similar study providing the mean and standard deviation, then the following formula can be applied for calculation of sample size for the contemplated study:

$$N = \frac{\left(Z_{1-\frac{a}{2}}\right)^2 * \sigma^2}{E^2} \qquad \text{(Formula 18.1)}$$

where E is an absolute error in same units as that of mean.

Exercise 18.1
An investigator wants to find the average height of paramedical college students. Calculate the sample size required if a pilot study revealed a mean height of 170 cm (SD = 10), taking errors as 1%.

Solution

$$\text{Taking} \left(Z_{1-\frac{a}{2}}\right) = 1.96$$

and

$$E = 170 * 1\% = 1.7$$

$$N = \frac{1.96^2 * 10^2}{1.7^2}$$

$$N = 132.9 \approx 133$$

Height of at least 133 students needs to be assessed.

18.3   Sample Size for Assessing Significance of Single Proportion

If, instead of the mean, the previous study has provided the proportion of the single variable intended to be studied in the investigator's planned study, the following method may be applied:

$$N = \frac{\left(Z_{1-\frac{a}{2}}\right)^2 * PQ}{E^2} \qquad \text{(Formula 18.2)}$$

where E is an absolute error in proportion P.

Exercise 18.2

If previous study revealed proportion of typhoid cases in a particular society as 12%, calculate the sample size to find the proportion of typhoid cases in same society with 10% error.

Solution

$$P = 0.12 \text{ and } Q = 1 - P = 0.88$$

$$\text{Taking } \left(Z_{1-\frac{\alpha}{2}}\right) = 1.96$$

and

$$E = 0.12 * 10\% = 0.012$$

$$N = \frac{1.96^2 * 0.12 * 0.88}{0.012^2}$$

$$N = 2818$$

Hence at least 2818 individuals need to be assessed to get meaningful result at desired error level.

18.4 Sample Size for Assessing Significance of Two Means

Similarly, when there is a previous similar study providing mean and SD of two groups (case and control or two different groups), the sample size required for each group can be calculated as

$$N \text{ (each group)} = \frac{\left(Z_{1-\frac{\alpha}{2}} + Z_{1-\beta}\right)^2 * (\sigma_1^2 + \sigma_2^2)}{(\overline{x_1} - \overline{x_2})^2} \qquad \text{(Formula 18.3)}$$

Exercise 18.3

An investigator wants to assess the difference in mean platelet count between megaloblastic anaemia and control group. Calculate the sample size required in each group taking confidence of 95% and power 80%, if previous study revealed mean platelet count of $125*10^3$ per mm^3(SD = 50) and mean platelet count of control group was $150*10^3$ per mm^3(SD = 20).

Solution

$$\overline{x_1} = 125$$

$$\sigma_1 = 50$$

$$\overline{x_2} = 150$$

$$\sigma_2 = 20$$

Taking $\left(z_{1-\frac{a}{2}}\right) = 1.96$ and $(z_1 - \beta) = 0.842$ (i.e. with 80% power of study)

$$N \text{ (each group)} = \frac{(1.96 + 0.8412)^2 * (50^2 + 20^2)}{(125 - 150)^2}$$

$$N \text{ (each group)} = 36$$

Hence, each group (control and cases) should have at least 36 subjects.

18.5   Sample Size for Assessing Significance of Two Proportions

Commonly, the investigator has results of a previous study giving proportions of one or more variables in two or more groups. These can be utilized for calculation of sample size in each group for his/her contemplated study as

$$N \text{ (each group)} = \frac{\left(Z_{1-\frac{a}{2}}\sqrt{\frac{(p_1+p_2)(q_1+q_2)}{2}} + Z_{1-\beta}\sqrt{p_1 q_1 + p_2 q_2}\right)^2}{(p_1 - p_2)^2} \qquad \text{(Formula 18.4)}$$

Exercise 18.4

If a previous study on breast cancer revealed Her2neu positivity of 45% in cases with lymph node metastasis and positivity of 40% in cases without lymph node metastasis, calculate the sample size in each group taking confidence of 95% and power 80%, for an investigator who wants to assess the difference in her2neu positivity.

Solution

$$p_1 = 0.45 \; q_1 = 0.55$$

$$p_2 = 0.4 \; q_2 = 0.6$$

$$N \text{ (each group)} = \frac{\left(Z_{1-\frac{a}{2}}\sqrt{\frac{(p_1+p_2)(q_1+q_2)}{2}} + Z_{1-\beta}\sqrt{p_1 q_1 + p_2 q_2}\right)^2}{(p_1 - p_2)^2}$$

$$N \text{ (each group)} = 1532$$

At least 1532 patients need to be assessed in each group.

18.6   Sample Size for Correlation Coefficient

Sample size calculation for the purpose of assessing the correlation between two variables in a population differs from the ones described so far. For this purpose, a correlation coefficient provided by a previous similar study is utilized:

$$N = \left(\frac{Z_{1-\frac{a}{2}} + Z_{1-\beta}}{Cr_1}\right)^2 + 3 \qquad \text{(Formula 18.5)}$$

where

$$Cr_1 = \frac{1}{2} \ln \left(\frac{1+r}{1-r}\right)$$

Exercise 18.5
An investigator wants to the assess correlation between BMI and serum total cholesterol. A previous study revealed a correlation of 0.65. Find the sample size required. Taking confidence of 95% and power 80%.

Solution

$$r = 0.65$$

$$Cr_1 = \frac{1}{2} \ln \left(\frac{1+r}{1-r}\right)$$

$$Cr_1 = \frac{1}{2} \ln \left(\frac{1+0.65}{1-0.65}\right)$$

$$Cr_1 = \frac{1}{2} \ln \left(\frac{1.65}{0.35}\right)$$

Using natural logarithmic table

$$Cr_1 = 0.78$$

Taking $\left(Z_{1-\frac{a}{2}}\right) = 1.96$ and $(Z_{1-\beta}) = 0.842$ (i.e. with 80% power of study)

$$N = \left(\frac{Z_{1-\frac{\alpha}{2}} + Z_{1-\beta}}{Cr_1}\right)^2 + 3$$

$$N = \left(\frac{1.96 + 0.84}{0.78}\right)^2 + 3$$

$$N = (3.6)^2 + 3$$

$$N = 15.96 \approx 16$$

Minimum sample size of 16 subjects is required to assess this correlation.

18.7   Sample Size for Sensitivity and Specificity

All screening or diagnostic tests need to be evaluated for their test characteristics, i.e. sensitivity and specificity for diagnosing the intended condition. If a study is to be conducted to calculate the sensitivity and specificity of a test or modality in a population, the following formula can be utilized:

$$N_{Sn} = \frac{\left(Z_{1-\frac{\alpha}{2}}\right)^2 * Sn * (1 - Sn)}{E^2 * Prev} \qquad \text{(Formula 18.6)}$$

$$N_{Sp} = \frac{\left(Z_{1-\frac{\alpha}{2}}\right)^2 * Sp * (1 - Sp)}{E^2 * (1 - Prev)} \qquad \text{(Formula 18.7)}$$

where *Prev* is prevalence of disease in fraction and E is an absolute error.

Exercise 18.6
An investigator wants to estimate the sensitivity of a particular test in diagnosing a disease. What should be the sample size if a previous study found its sensitivity of 85% and the disease prevalence of 0.35 with an acceptable error of 7%?

Solution

$$Sn = 0.85, Prev = 0.35, E = 0.07$$

$$N_{Sn} = \frac{\left(Z_{1-\frac{\alpha}{2}}\right)^2 * Sn * (1 - Sn)}{E^2 * Prev}$$

$$N_{Sn} = \frac{1.96^2 * 0.85 * 0.15}{0.07^2 * 0.35}$$

$$N_{Sn} = 285.48 \approx 286$$

Hence, at least 286 patients need to studied to assess the sensitivity for the given error level.

Exercise 18.7
An investigator wants to estimate the specificity of a particular test. What should be the sample size if previous study found its specificity of 87% and disease prevalence was 0.35 with an acceptable error of 7%?

Solution

$$Sp = 0.87, Prev = 0.35, E = 0.07$$

$$N_{Sp} = \frac{\left(Z_{1-\frac{\alpha}{2}}\right)^2 * Sp * (1 - Sp)}{E^2 * (1 - Prev)}$$

$$N_{Sp} = \frac{1.96^2 * 0.87 * 0.13}{0.07^2 * 0.65}$$

$$N_{Sp} = 135.35 \approx 137$$

Hence, at least 137 patients need to studied to assess the specificity for the given error level.

18.8 Sample Size for Univariate Logistic Regression (See Chap. 28)

In cases where a variable is going to be assessed for its effect on the outcome using logistic regression, the sample size can be calculated as

$$\Delta = \frac{1 + (1 + r^2)e^{5r^2/4}}{1 + e^{-r^2/4}}$$

$$N = \frac{\left(Z_{1-\frac{\alpha}{2}} + \left(e^{-r^2/4}\right)Z_{1-\beta}\right)^2}{BP * r^2}(1 + 2BP * \Delta) \qquad \text{(Formula 18.8)}$$

where BP is baseline probability.

18.9 Relative Risk

Relative risk is a ratio of the probability of an event occurring in the group exposed to a risk factor versus the probability of the same event occurring in the nonexposed group. If the relative risk of an outcome depending on the exposure to a certain factor is to be calculated from a sample, the size of such a sample may be calculated using the given formula:

$$N = \frac{\left(Z_{1-\frac{q}{2}} + Z_{1-\beta}\right)^2}{2\left(\sqrt{RR} - 2\right)^2} * \frac{RR}{PU}$$

(Formula 18.9)

where PU is equal to prevalence in unexposed group and RR is the relative risk defined by an available previous study on the topic.

18.10 Sample Size to Assess Significance of Difference in Two Incidences in a Cohort Study

If an investigator wants to study the significance of difference in two incidences of a particular event/disease, the sample size for such a study can be estimated as follows depending on whether the data is censored or not censored.

(a) For censored data, the known incidences are modified as follows:

$$f(\lambda) = \frac{\lambda^3 T}{\left(\lambda T - 1 + exp^{-\lambda T}\right)}$$

where T is duration of the censored study and λ is the incidence.

(b) For uncensored data $f(\lambda) = \lambda$.

For both the censored and uncensored situations, the sample size can be calculated as

$$n_1 = \frac{\left\{Z_{1-\frac{q}{2}}\sqrt{\left[(1+\kappa)f\left(\overline{\lambda}\right)\right]} + Z_{1-\beta}\sqrt{\kappa f(\lambda_1) + f(\lambda_2)}\right\}^2}{\left(\lambda_1 - \lambda_2\right)^2}$$

(Formula 18.10)

where $\overline{\lambda} = \frac{\lambda_1 + \lambda_2}{2}$, $\kappa = \frac{n_2}{n_1}$, $n_2 =$ control group size.

Exercise 18.8

An investigator wants to find incidence of a particular disease in a cohort study. He found a previous study with incidence of disease among exposed group of 35% and in unexposed group of 10%. Calculate the sample size.

Solution

$$\lambda_1 = 0.35, \lambda_2 = 0.1, \bar{\lambda} = 0.225 \quad \text{for } k = 1$$

$$n_1 = \frac{\left\{1.96\sqrt{[(1+1)\,0.225]} + 0.84\sqrt{1 * 0.35 + 0.1}\right\}^2}{(0.35 - 0.1)^2}$$

$$n_1 = \frac{\left\{1.96\sqrt{0.45} + 0.84\sqrt{0.45}\right\}^2}{(0.25)^2}$$

$$n_1 = \frac{\left\{1.313 + 0.562\right\}^2}{(0.25)^2}$$

$$n_1 = \frac{3.515}{0.0625}$$

$$n_1 = 56.25 \approx 57$$

$$n_2 = k * n_1 = 57$$

Calculated sample size is 57 cases in each of the exposed and nonexposed group.

18.11   Sample Size for Case Control and Cohort Studies

Epidemiological studies, including case control and cohort, are frequently undertaken to assess the significance of one or more factors in causation or prognosis of diseases. For these studies to provide meaningful results, the sample size calculation should be done diligently before the study is undertaken. The following formula can be used for such a calculation:

$$n_1 = \frac{(Z_{1-\frac{\alpha}{2}} + Z_{1-\beta})^2 \bar{p}\,\bar{q}(r+1)}{r(p_1 - p_2)^2} \qquad \text{(Formula 18.11)}$$

where

$$\bar{p} = \frac{p_1 + rp_2}{r+1}$$

$$\bar{q} = 1 - \bar{p}$$

n_1 = size of study group,
n_2 = size of control group,
$r = n_1/n_2$,
p_1 = proportion of cases with exposure for case control study or exposed group in cohort,
p_2 = proportion of controls with exposure for case control study or unexposed group in cohort.

If instead of proportion (p_1), odds ratio (OR) is known from previous study, then

$$p_1 = \frac{p_2 OR}{1 + p_2(OR - 1)}$$

If instead of proportion (p_1), relative risk (RR) is known from previous study, then

$$p_1 = p_2 RR$$

Exercise 18.9
An investigator wants to estimate the relative risk of a particular disease. He found a previous study with proportion of disease among nonexposed group of 35% and a relative risk of 0.5. Find out the sample size with confidence of 95% and power 80%.

Solution

$$RR = 0.5, p_2 = 0.35, p_1 = 0.175, \quad \text{for } r = 1$$

$$\bar{p} = \frac{p_1 + rp_2}{r + 1} = \frac{0.175 + 1 * 0.35}{1 + 1}$$

$$\bar{p} = 0.26$$

$$\bar{q} = 1 - \bar{p}$$

$$\bar{q} = 1 - 0.26$$

$$\bar{q} = 0.74$$

$$n_1 = \frac{(Z_{1-\frac{q}{2}} + Z_{1-\beta})^2 \bar{p}\,\bar{q}(r + 1)}{r(p_1 - p_2)^2}$$

$$n_1 = \frac{(1.96 + 0.84)^2 0.26 * 0.74(1 + 1)}{1(0.175 - 0.35)^2}$$

$$n_1 = \frac{(2.8)^2 * 0.384}{(-0.175)^2}$$

$$n_1 = \frac{7.84 * 0.384}{0.0306}$$

$$n_1 = 98.3 \approx 99$$

Hence, 99 subjects in exposed group and 99 subjects in unexposed group will be needed for the study.

18.12 Sample Size for Unequal Samples

$$N_u = \frac{N_f * N_{cal}}{2N_f - N_{cal}} \qquad \text{(Formula 18.12)}$$

where

N_{cal} is calculated N by using any of the formula describe in this chapter considering equal number cases in two groups.
N_f is fixed group size and N_u is size of unfixed group.

Exercise 18.10
If in a situation given in exercise 18.4, the number of cases without lymph node metastasis is limited to 1000. How many cases with lymph node metastasis to be taken for same confidence and power?

Solution

$$N_f = 1000, N_{cal} = 1532$$

$$N_u = \frac{1000 * 1532}{2 * 1000 - 1532}$$

$$N_u = \frac{1532000}{468}$$

$$N_u = 3273.5 \approx 3274$$

About 3274 breast cancer patients with lymph node metastasis will be required.

18.13 Sample Size for Finite Population

In cases where the population from which the sample is to be drawn is a finite one, i.e. the exact size of the population is known, the formula for calculation of sample size is modified to

$$N = \frac{N_{cal}}{1 + (N_{cal} - 1)/Pop} \qquad \text{(Formula 18.13)}$$

where *Pop* is population size and N_{cal} is calculated sample size derived from one or more formulas described for sample size estimation in this chapter.

Exercise 18.11

If in a situation given in exercise 18.2, total population of the society is 5000. Calculate the sample size for finite population.

Solution

$$N_{cal} = 2818, Pop = 5000$$

$$N = \frac{N_{cal}}{1 + (N_{cal} - 1)/Pop}$$

$$N = \frac{2818}{1 + (2818 - 1)/5000}$$

$$N = \frac{2818}{1 + 2817/5000}$$

Sample size required for finite population of 5000 is 1803.

Epidemiological Studies

19

Abstract

Background
Cohort study
Case control Study
Cross-sectional Study
Experimental study
Sources of errors in epidemiological studies
Designing a study and writing a study report
Steps in designing a study
Writing a study report

Keywords

Epidemiological study · Cohort study · Case control study · Cross-sectional study · Experimental study · Study report

19.1 Background

Nearly all facts and evidences in medicine are collected through epidemiological studies. There are many types of epidemiological studies, and the difference between these should be amply clear in the minds of all students and researchers. The epidemiological studies can be broadly classified into observational studies and experimental studies, the latter being randomized controlled trials. Each of these can be further subclassified as shown in Fig. 19.1.

Descriptive studies usually result in hypothesis generation, whereas analytical studies test the hypothesis thus generated. Now let us look at some of the most common studies undertaken in the medical science.

© Springer Nature Singapore Pte Ltd. 2019
S. K. Yadav et al., *Biomedical Statistics*,
https://doi.org/10.1007/978-981-32-9294-9_19

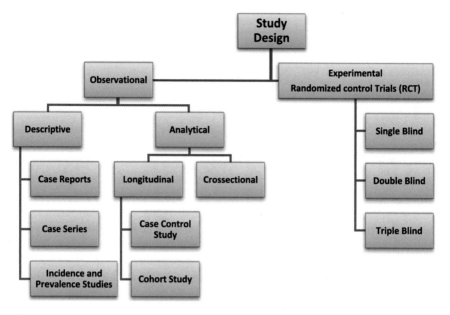

Fig. 19.1 Classification of study design

19.2 Cohort Study

Cohort study begins with an exposure (risk factor) and follows it to the outcome. Although cohort studies are usually prospective, they may be concurrent or retrospective as well (Fig. 19.2). A robust well-designed cohort study usually provides the best evidence of the association of a risk factor with causation of a disease.

The word *cohort* was used for ancient roman military unit of 300–600 men marching forward in a battle. In epidemiological studies, cohort is used for group of persons having exposure to a particular situation.

Example (Table 19.1)

Research Question (Table 19.2) Does diabetes mellitus increase the risk of developing tuberculosis?

Incidence of tuberculosis with diabetes present $= \frac{7}{500} * 100 = 1.4$.
Incidence of tuberculosis with diabetes absent $= \frac{2}{1000} * 100 = 0.2$
Relative risk $= \frac{\text{Incidence } 1}{\text{Incidence } 2} = 7.0$

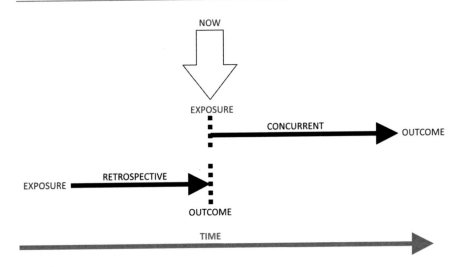

Fig. 19.2 Concurrent and retrospective cohort study design

Table 19.1 Cohort of patients with diabetes mellitus and normal subjects followed up for 2 years

	Population follow-up 2 years	Cases of tuberculosis
Diabetes present	500	7
Diabetes absent	1000	2

Table 19.2 Incidence and relative risk of cohort shown in Table 19.1

	Population follow-up 2 years	Cases of TB	Incidence (%)	Relative risk
Diabetes present	500	7	1.4	7.0
Diabetes absent	1000	2	0.2	

This study shows that diabetic patients have seven times increased risk of developing tuberculosis.

19.2.1 Advantages

1. Incidence and relative risk of a disease can be accurately estimated.
2. Since it starts with the exposed individuals, rare exposure can also be studied.
3. Multiple outcomes of a particular exposure can be studied at the same time since the study starts from exposure and follows up to the outcome.

4. Temporal relationships between exposure and outcome and between multiple outcomes can be inferred.

However, there are certain disadvantages of conducting a cohort study, as mentioned below.

19.2.2 Disadvantages

1. Since certain exposures may result in outcome after a long period of time, cohort studies are usually lengthy and expensive to conduct. Hence, it is not easy for investigators to conduct a well-designed cohort study in many situations.
2. It is not suitable for rare diseases, since the exposure may not be known.
3. It is not suitable for diseases with long latency since the time from exposure to outcome may be unduly long and impractical for a follow-up study.
4. During follow-up, following problems may arise:
   (a) Migration of the study population.
   (b) Loss of follow-up.
   (c) Additional environmental changes influencing the outcome. Hence, the results of a cohort study need to be presented and evaluated in light of these phenomena.

19.3   Case Control Study

In contrast to cohort study, a case control study begins with outcome and traces the exposure. As the name suggests, case control studies have two groups as follows:

- Case: individuals with a particular outcome.
- Controls: individuals without the particular outcome.

These two groups are then questioned for the presence of a particular exposure in their life history.

Example (Table 19.3)

Research Question: Is smoking associated with lung cancer?

Table 19.3 Cases and control to study the association of smoking and lung cancer

	Lung cancer cases	Controls
Smokers	42 (a)	45 (b)
Non smokers	25 (c)	120 (d)

$$\text{Odds Ratio (OR)} = \frac{\left(\frac{a}{c}\right)}{\left(\frac{b}{d}\right)} = \frac{ad}{bc}$$

$$\text{OR} = \frac{42 * 120}{45 * 25}$$

$$\text{OR} = 4.48$$

This study shows that odds are in favour of association of smoking and lung cancer.

19.3.1 Advantages

1. Case control studies are much quicker and cheaper to conduct than cohort studies.
2. Rare diseases can be studied, since the study begins with the outcome.
3. It is also suitable for disease with long latency for the same reason.
4. Multiple exposures can be studied for the same outcome.

However, case control studies are fraught with few disadvantages, because of which the evidence provided by these studies is considered to be inferior to those given by cohort studies.

19.3.2 Disadvantages

1. Incidence estimation is not possible since the study begins with the outcome. Hence, the incidence in an exposed population cannot be estimated with confidence.
2. Multiple outcomes cannot be studied.
3. Due to problem of recall, it may be difficult to infer the temporal relationship confidently.

19.4   Cross-Sectional Study

In cross-sectional study, all the observations or measurements are made at a cross-section of time (i.e. a point of time). Although this type of study design is mostly exploratory or descriptive in nature, tests described in inferential statistics in this book are sometimes useful to analyse the results of such a study. An example of cross-sectional study includes a study involving presence or absence of tumour markers for a particular tumour in cases and controls.

Table 19.4 Types of blinded experimental studies

Type of study	To be blinded		
	Party 1	Party 2	Party 3
Single blinded	Study subject	–	–
Double blinded	Study subject	Individual who administers treatment	–
Triple blinded	Study subject	Individual who administers treatment	Individual who measures outcome

19.5 Experimental Study

An experimental study involves two groups. In one group known as the experimental group, some interference or intervention is done. An example of intervention is treatment of a certain disease. In the other group, no interference is done, and this is known as the control group. Variable(s) of interest is measured in both the groups, and tests for inferential statistics, as applicable, are applied. Control and experimental groups must be matched with respect to all potentially confounding variables except for the intervention proposed.

Sometimes, there is only one group, and variable is measured before and after treatment to get paired observations.

To reduce bias in an experimental study, blinded trials are preferred. Blinded trials may be single blinded, double blinded or triple blinded depending on the number of parties who are blinded to the relevant information, as shown in Table 19.4.

19.6 Sources of Errors in Epidemiological Studies

The epidemiological studies, both cohort and case control, are subject to a few sources of error which have to be kept in mind during their designing as well as interpretation of their results.

19.6.1 Random Error

Random error results due to inherently unpredictable fluctuations in a particular sample/population. It can be reduced by a large sample size (hence the importance of calculation of sample size) and by replication methods, if possible.

19.6.2 Bias

Bias is the intentional or unintentional favouring of one group or outcome over other potential groups or outcomes in the population. It is reduced by being aware and careful of the existence of this source of error.

19.6.3 Confounding

Confounder, also called as confounding factor, is a variable which influences both the dependent and independent variables in the study leading to spurious association. The error induced by confounders can be reduced by randomization, restriction of confounder, stratification by confounders and multivariate analysis.

19.6.4 Effect Modification

If outcome is also dependent on a third variable (other than the independent variable in the study), then it is known as effect modification. It can be reduced by stratification and multivariate analysis.

19.6.5 Reverse Causation

Reverse causation may be due to either reverse cause-and-effect relationship or two-way causal relationship. Reverse causation occurs when subjects change their behaviour (exposure) after developing a disease (outcome). It is assessed by Mendelian randomization.

19.7   Designing a Study and Writing a Study Report

Before embarking on a study, students and researcher should appraise themselves with the various steps involved in the designing of a particular study and in preparation of its report.

19.7.1 Steps in Designing a Study

1. Problem definition – define the problem or disease or exposure to be studied.
2. Aim and objective – these should be clear, unambiguous and realistic.
3. Review of literature – literature review should be thorough providing the basis of conduct of the proposed study.

4. Formulation of hypothesis – the alternate hypothesis and null hypothesis should be formulated at this step.
5. Designing of methodology – methodology should be commensurate with the aims and objective of study and should be feasible within the framework of the study. Population is defined, sample size is calculated and sampling method is decided at this step.
6. Recording and presentation of data – the method of data recording and presentation as table or graphs is usually considered at the initiation of study and refined later.
7. Statistical analysis – once data is presented appropriately, the most suitable statistical method can be judged and applied for meaningful analysis.
8. Conclusion and extrapolation – based on the results of statistical analysis, conclusions can be drawn rejecting or accepting the null hypothesis. Depending on the population characteristics and sampling method employed, extrapolation of the study results may be attempted. The limitations, if any, of the study should also be considered while suggesting the extrapolation.

19.7.2 Writing a Study Report

Writing and presenting the study report is as important, if not more, as the method of conducting it. The study report should essentially include the following sections:

1. Title
2. Abstract
3. Introduction
4. Review of literature
5. Materials and methods
6. Results and observations
7. Discussion
8. Conclusion
9. Acknowledgement
10. References
11. Appendices

Analysis of Diagnostic Test

20

Abstract

Background
Sensitivity
Specificity
Positive predictive value
Negative predictive value
Accuracy
Receiver operating characteristic curve analysis
Cohen's Kappa Bland and Altman plot

Keywords

Sensitivity · Specificity · Positive predictive value · Negative predictive value · Accuracy · ROC curve · Cohen's Kappa · Bland and Altman plot

20.1 Background

Laboratory tests can be broadly classified into screening tests and diagnostic tests. Screening tests are conducted in apparently healthy individuals to detect the earliest signs of a disease, like cancer. On the other hand, diagnostic tests are performed for confirmation or exclusion of the clinical suspicion of a disease in a symptomatic patient. All such tests, before being implemented in the health practice, are subjected to analysis of usability through the following parameters:

1. Sensitivity
2. Specificity
3. Positive predictive value
4. Negative predictive value
5. Accuracy

© Springer Nature Singapore Pte Ltd. 2019
S. K. Yadav et al., *Biomedical Statistics*,
https://doi.org/10.1007/978-981-32-9294-9_20

Table 20.1 The definitions used in formulas in this chapter

Test result	Disease		Total
	Present	Absent	
Positive	True positives (A)	False positive (B)	A + B
Negative	False negative (C)	True negative (D)	C + D
Total	A + C	B + D	A + B + C + D

For calculation of these parameters for any test, the definitions used are given in Table 20.1.

True positives are those that have the disease and are tested as positive by the screening/diagnostic test. Similarly, true negatives are the ones that are correctly identified as not having the disease being detected by the test. On the other hand, false positives are those that are tested positive by the test but do not have the disease in reality. While false negative are the ones with the disease that are not detected by the test.

Now, let us proceed with understanding the calculation of these various parameters.

20.2 Sensitivity (Sn)

Sensitivity is a measure of proportion of true positives among diseased individuals.

$$Sn = \frac{\text{Number of true positives}}{\text{Number of patients having disease}}$$

$$Sn = \frac{\text{Number of true positives}}{\text{Number of true positives} + \text{Number of false negatives}}$$

$$Sn = \frac{A}{A + C}$$

20.3 Specificity (Sp)

Specificity is a measure of proportion of true negatives among the healthy individuals under study.

$$Sp = \frac{\text{Number of true negatives}}{\text{Number of healthy subjects}}$$

$$Sp = \frac{\text{Number of true negatives}}{\text{Number of true negatives} + \text{Number of false positives}}$$

$$Sp = \frac{D}{B + D}$$

20.4   Positive Predictive Value (PPV)

It is the probability of having the disease, given that the individual has a positive test result. It is the proportion of true positives out of total positives given by the diagnostic test. The ideal PPV is 1 (100%) for a diagnostic test to be called "perfect".

$$PPV = \frac{Number\ of\ true\ positives}{Number\ of\ all\ tested\ positive}$$

$$PPV = \frac{Number\ of\ true\ positives}{Number\ of\ true\ positive + number\ of\ false\ positives}$$

$$PPV = \frac{A}{A + B}$$

PPV can also be calculated from sensitivity, specificity and prevalence by using the given formula:

$$PPV = \frac{sensitivity * prevalence}{sensitivity * prevalence + (1 - specificity) * (1 - prevalence)}$$

20.5   Negative Predictive Value (NPV)

It is the probability of an individual not having the disease, given that he/she has a negative test result. It is the proportion of true negatives out of total negatives given by the diagnostic test. The ideal NPV is 1 (100%) for a diagnostic test to be called perfect for exclusion of a disease.

$$NPV = \frac{Number\ of\ true\ negatives}{Number\ of\ all\ tested\ negative}$$

$$NPV = \frac{Number\ of\ true\ negatives}{Number\ of\ true\ negatives + number\ of\ false\ negatives}$$

$$NPV = \frac{D}{C + D}$$

Similar to the PPV, NPV can also be calculated from sensitivity, specificity and prevalence by using the given formula:

$$\text{NPV} = \frac{\text{specificity} * (1 - \text{prevalence})}{(1 - \text{sensitivity}) * \text{prevalence} + \text{specificity} * (1 - \text{prevalence})}$$

Hence, as we observe, both PPV and NPV are affected by the prevalence of the disease in the population where the diagnostic test is going to be implemented.

20.6   Accuracy

It is the proportion of all true result (positive and negative) out of the total tests performed.

$$\text{Accuracy} = \frac{\text{Number of true positives} + \text{number of true negatives}}{\text{Total number of tests performed}}$$

$$\text{Accuracy} = \frac{A + D}{A + B + C + D.}$$

Exercise 20.1
Calculate the sensitivity, specificity, positive predictive value, negative predictive value and accuracy of a screening test to detect colon cancer. Data shown in Table 20.2.

Solution (Table 20.3)

$$\text{Sensitivity} = \frac{A}{A + C} = \frac{90}{95} = 0.947 \text{ i.e.} 94.7\%$$

Table 20.2 Data of test result for diagnosis of cancer

Test result	Cancer present	Cancer absent
Positive	90	10
Negative	5	1395

Table 20.3 Labelling of data for calculation

Test result	Cancer present	Cancer absent	Total
Positive	$A = 90$	$B = 10$	$A + B = 100$
Negative	$C = 5$	$D = 1395$	$C + D = 1400$
Total	$A + C = 95$	$B + D = 1405$	$A + B + C + D = 1500$

$$\text{Specificity} = \frac{D}{B+D} = \frac{1395}{1405} = 0.992 \text{ i.e.} 99.2\%$$

$$\text{PPV} = \frac{A}{A+B} = \frac{90}{100} = 0.90 \text{ i.e.} 90\%$$

$$\text{NPV} = \frac{D}{C+D} = \frac{1395}{1400} = 0.996 \text{ i.e.} 99.6\%$$

$$\text{Accuracy} = \frac{A+D}{A+B+C+D} = \frac{90+1395}{1500} = 0.99 \text{ i.e.} 99\%$$

20.7   Receiver Operating Characteristic Curve (ROC) Analysis

This technique is used to evaluate a diagnostic test for a disease or to calculate the appropriate cut-off for screening or diagnostic test. It is derived from the work done during World War II to identify radar signals for the prediction of outcome of interest. For ROC analysis, initially sensitivity and specificity are calculated. It is an iterative process done with various cut-off points to identify (diagnose) the outcome of interest. In ROC analysis sensitivity and specificity are calculated for various cut-offs, and a graph is plotted between sensitivity and 1-specificity. The more the graph is away from 45° diagonal, the better the test is (see Fig. 20.1).

To identify the appropriate cut-off, sensitivity and specificity are plotted in a graph against the results of diagnostic test. The best cut-off is the corresponding point on x-axis where both the curves intersect as shown in Fig. 20.2.

20.8   Measures of Agreement

20.8.1 For Qualitative Data: Inter-Rater Agreement (Cohen's Kappa Statistics)

Kappa statistics is a measure of the agreement between two observers for rating or classification of the total number of items in a study into two or more mutually exclusive categories. If P_e is the hypothetical possibility of chance agreement between the two observers and P_a is the observed agreement, then kappa is given as

$$\kappa = \frac{P_a - P_e}{1 - P_e}$$

A simple case of 2×2 table calculation of P_e and P_a is illustrated in Table 20.4.

$$P_e = \left[\frac{C_1}{T} * \frac{R_1}{T}\right] + \left[\frac{C_2}{T} * \frac{R_2}{T}\right]$$

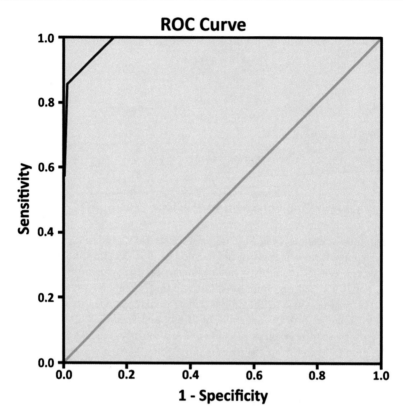

Fig. 20.1 An ROC curve (blue) comfortably farther than diagonal line (green)

$$P_a = \frac{A+D}{T}$$

For interpretation of kappa statistics, the most frequently used categorization is hown in Table 20.5.

In biomedical field, especially pertaining to cancer screening or diagnosis, weighted kappa is used. In this method, weights are ascribed to the various diagnostic categories for consideration of interobserver agreement or disagreement, and these weights are then utilized for calculation of kappa statistics.

Exercise 20.2
Two pathologists examined the breast biopsy slides of 70 patients and classified them into benign and malignant categories. The data is given in Table 20.6.

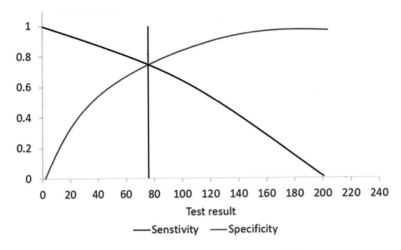

Fig. 20.2 Determination of cut-off point using sensitivity and specificity curve

Table 20.4 Scheme for calculation of hypothetical possibility (P_e) and observed agreement (P_a)

	Observer 1		
Observer 2	Category 1	Category 2	Total
Category 1	A	B	R_1
Category 2	C	D	R_2
Total	C_1	C_2	T

Table 20.5 Interpretation of kappa

Kappa value	Interpretation
<0	No agreement
0.0–0.2	Slight agreement
0.21–0.4	Fair agreement
0.41–0.6	Moderate agreement
0.61–0.8	Substantial agreement
0.81–1.00	Almost perfect agreement

Table 20.6 Results of categorizations by two pathologists

	Pathologist 1		
Pathologist 2	Benign	Malignant	Total
Benign	25	5	30
Malignant	6	34	40
Total	31	39	70

Solution

$$P_e = \left[\frac{C_1}{T} * \frac{R_1}{T} \right] + \left[\frac{C_2}{T} * \frac{R_2}{T} \right]$$

$$P_e = \left[\frac{31}{70} * \frac{30}{70}\right] + \left[\frac{39}{70} * \frac{40}{70}\right]$$

$$P_e = 0.508$$

$$P_a = \frac{A + D}{T}$$

$$P_a = \frac{25 + 34}{70}$$

$$P_a = 0.842$$

$$\kappa = \frac{P_a - P_e}{1 - P_e}$$

$$\kappa = \frac{0.842 - 0.508}{1 - 0.508}$$

$$\kappa = 0.678$$

There is substantial agreement between the two pathologists.

20.8.2 For Quantitative Data: Bland and Altman Plot

For the purpose of determination of agreement between two analytical methods providing quantitative results, Bland and Altman, in 1986, proposed the use of a *difference plot*. In this, the graph is prepared by plotting the paired differences in values provided by the two methods on y-axis and the means of values of each pair on the x-axis. The agreement between the methods is assessed from the closeness of the plotted points to the "nil bias" or "zero-difference" line. The closer the points are to this line, the higher is the agreement between the two methods for a given estimation. The Bland and Altman plot is quite useful in a clinical laboratory for comparison of two methods for biochemical estimation.

Mean bias is given as

$$\bar{d} = \frac{1}{n} \sum d_i$$

Standard deviation is

$$S_d = \sqrt{\frac{1}{n-1} \sum \left(d_i - \bar{d}\right)^2}$$

95% confidence intervals of mean bias are calculated as

$$\overline{d} \pm t_{1-\alpha/2,\, n-1} * \left(\frac{S_d}{\sqrt{n}}\right)$$

Exercise 20.3
In a particular laboratory, blood glucose levels of a set of 11 subjects were measured using two different analytical methods. The results of which are shown in Table 20.7. Assess the agreement between two methods and illustrate using a suitable diagram.

Solution
The paired differences between two methods are tabulated in Table 20.8 and Fig. 20.3.

Table 20.7 Blood glucose levels (gm/dL) resulting from two different methods

Method A	Method B
95	93
105	101
110	115
75	78
81	77
76	78
114	110
103	105
108	110
110	106
115	110

Table 20.8 Paired differences of data in Table 20.6

Method A	Method B	Mean	Difference (d_i)
95	93	94	2
105	101	103	4
110	115	112.5	−5
75	78	76.5	−3
81	77	79	4
76	78	77	−2
114	110	112	4
103	105	104	−2
108	110	109	−2
110	106	108	4
115	110	112.5	5

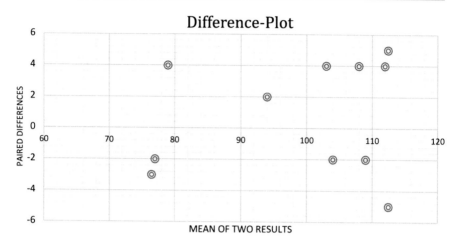

Fig. 20.3 Difference plot of the data given in exercise 20.3

$$\bar{d} = \frac{1}{n} \sum d_i$$

$$\bar{d} = \frac{1}{11} * 9$$

Mean Bias $\bar{d} = 0.818$

Part III

Demography and Life Tables

Demography

21

Abstract

Demography
Utility of demographic data
Sources of demographic data
Stages of life

Keywords

Demographic data · Census · Vital statistics

21.1 Demography

Demography is the study of size, composition and distribution of a population. This branch of epidemiology constitutes three types:

Static demography is a study of a population at a given point in time.
Dynamic demography is a study of changes in a population over time.
Vital statistics means quantitative data of vital events like birth and death in life (with regard to the population).

21.2 Utility of Demographic Data

Demographic data of a population is very important and useful for a number of assessments and planning at the national or state level, such as:

1. For assessment of the national needs for essential items (e.g. food and other necessities).

2. Planning of health and welfare programs at the state and national levels.
3. Monitoring of health and welfare programs after their implementation.
4. Comparison of fatality, mortality and morbidity of different communities, professions, age groups, genders and socioeconomic groups.
5. For calculation of life expectancy, which is extremely useful in insurance business and to compare the economic status of a country against the world.

Given the above uses of demographic data, collection of such data using reliable tools and data sources becomes imperative since once collected, the data is used by multiple agencies.

Demographic Publications in India Are Available from

1. World Health Organization
2. Registrar General of India
3. Directorate General of Health Services (DGHS), India
4. State health directorates
5. National and State Cancer Registries

21.3 Sources of Demographic Data

A number of sources are utilized in collection of demographic data from any population. In India, these sources include:

1. Population census
2. Record of vital statistics
3. Records from health institutions
4. Records of special surveys

21.3.1 Population Census

It is the process of collection, compilation and publication of demographic, economic and social data at specific time in a specific territory. It is collected from each and every individual residing in the given territory. The first census report of India was released in 1872 followed by 1881. Thereafter, census in India has been done every 10 years under the Indian Census Act 1948. The process of census consists of the following steps and phases:

1. Planning according to the "Principles and recommendations for National Population Census" given by the United Nations.
2. Wide publicity of the process to ensure maximum inclusion and cooperation.
3. Pilot study.

4. Actual census.
5. Publication of census data.

21.3.2 Record of Vital Statistics

Recording of vital statistics in India is done under the following acts:

1. Bengal Birth and Death Registration Act, 1873.
2. Registration of Births and Death Act, 1969: As per this act, registration of births and deaths is the responsibility of the state. For this purpose, the state governments appoint District Registrars and State Chief Registrars who maintain these vital registers. Registration of a vital event is mandatory within 15 days of birth and 7 days of death of an individual in a family residing in a particular state.
3. Sample Registration System (SRS), 1964: SRS records birth, death, fertility, etc. by random stratified sampling technique.
4. Model Registration System, 1965: This has been renamed as Survey of Causes of Death in 1982. It uses the stratified sampling method and is done with the help of trained paramedical person of primary health centre.

21.3.3 Records of Health Institutions

Monthly reports from health institutions are sent to the state directorates. These reports are meant to contain all statistics including the notifiable diseases. Though this is a useful data source, the data from hospitals is usually biased because it does not cover any specific area of population and a certain hospital in any given region may cater to patients from other regions as well. However, these records can amply be used to assess the hospital needs.

21.3.4 Records of Special Surveys

Special surveys are done by an epidemiologist and health expert. These are done for special purposes like the estimation of the prevalence of chronic diseases like tuberculosis, leprosy, cancer, etc. and for assessment of nutritional status.

21.4   Stages of Life

For the purpose of collection of demographic data, the stages of life are described in following terms and practiced uniformly (Table 21.1).

Table 21.1 Stages of life

Stage	Period of life
Embryo	Conception to 8 weeks of gestation
Fetus	9–28 weeks of gestation
Perinatal period	28 weeks of gestation to 7 days after birth
Neonate	Birth to 28 days
Early neonate	Birth to 7 days
Late neonate	8–28 days
Infant	Birth to 1 year
Toddler	1–3 years
Preschool	2–5 years
Adolescent	13–19 years

Table 21.2 Definitions of vital events

Vital event	Definition
Fetal death	<28 weeks of gestation
Early fetal death	<20 weeks of gestation
Intermediate fetal death	Between 20 and 28 weeks of gestation
Still birth	Late fetal death >28 weeks of gestation
Prematurity	Birth weight < 2500 grams
Infant death	<1 year
Neonatal death	<28 days
Perinatal death	28 weeks of gestation to <7 days
Maternal death	Death due to complication of pregnancy, child birth or puerperium

Uniform application of these life stages ensures comparability between data from different regions, states, countries and continents.

As mentioned earlier, let us also understand the definitions of vital events recorded in the demographic data (Table 21.2).

Measures of Demography

22

Abstract

Measures of population
   Mid-year population
      Natural increase method
      Arithmetic progression method
      Geometric progression method
   Population density
Measures of fertility and reproduction
   Crude birth rate
   Fertility rate
      General fertility rate
      Age specific fertility rate
      Total fertility rate
   Reproduction rate
      Gross reproduction rate
      Net reproduction rate
   Sex ratio at birth
   Illegitimacy rate
Measures of morbidity
   Disease incidence rate
   Disease prevalence rate
Measures of mortality
   Crude death rate
   Standardized death rate
   Specific death rate
      Age specific death rate
      Infant mortality rate
      Fetal death ratio
      Still birth rate or late fetal death rate
      Perinatal mortality rate

© Springer Nature Singapore Pte Ltd. 2019
S. K. Yadav et al., *Biomedical Statistics*,
https://doi.org/10.1007/978-981-32-9294-9_22

Neonatal mortality rate
Post neonatal mortality rate
Maternal mortality rate
Sex specific death rate
Age and sex specific death rate
Cause specific death rate
Proportional mortality rate
Proportional mortality indicator (50 years)
Case fatality rate
Measures related to population control
Couple protection rate
Pearl index
Measures of health

Keywords
Measures of population · Measures of fertility · Measures of morbidity · Measures of mortality · Measures of health

22.1   Background

Demography of a population can be studied under the following headings:

1. Measures of population
2. Measures of fertility and reproduction
3. Measures of morbidity
4. Measures of mortality
5. Measures related to population control
6. Measures of health

22.2   Measures of Population

The actual demographic measure of population is population census. However, at times, inter-census (mid-year) population has to be estimated, and this can be done through one of the following methods:

22.2.1 Mid-year Population

22.2.1.1 Natural Increase Method
In this method, the addition to a population due to birth and immigration and depletion due to death and emigration are adjusted in the baseline population by the formula:

$$P_{t1} = P_{t0} + \text{(births and imigration)} - \text{(deaths and emigration)}$$

where births, deaths, immigration and emigration are counted during the period $t_1 - t_0$.

22.2.1.2 Arithmetic Progression Method

The baseline population is adjusted using an arithmetic rate. The arithmetic rate of increase per year is estimated from the previous inter-census period.

$$P_{t1} = P_{t0} + r * t$$

where $t = t_1 - t_0$, and r is the rate of increase per year as estimated from previous inter-census period.

Exercise 22.1

According to a census, the population of the state of Madhya Pradesh was 6.03 crore on 1 January 2001 and 7.27 crore on 1 January 2011. Calculate the mid-year population in 2016.

Solution

$$P_{t0} = 7.27$$

$$t = 5.5 \text{ year } (1 \text{ January } 2011 \text{ to } 30 \text{ June } 2016)$$

$$r = (7.27 - 6.03)/10 = 0.124$$

$$P_{t1} = 7.27 + 0.124 * 5.5 = 7.95$$

Hence, mid-year population on 30 June 2016 will be 7.95 crore, using the arithmetic progression method.

22.2.1.3 Geometric Progression Method

This method assumes that the increase in population over time does not follow an arithmetic rate. The baseline population is adjusted using geometric rate, which is estimated from previous inter-census period.

$$P_{t1} = P_{t0}(1 + r)^t$$

where $t = t_1 - t_0$

$$\log P_{t1} = \log P_{t0} + t \log (1 + r)$$

r is the rate of increase per year as estimated from previous inter-census period after rearranging the above equation.

$$\log\left(1 + r\right) = \frac{\log P_{t0} - \log P_{t-1}}{i}$$

here i is inter-census period, t_0 is recent census and $t-1$ is previous census.

22.2.2 Population Density

Density of population is the population in numbers of a geographic location divided by the area of that geographic location. This measure is useful in demographic comparison of different areas by providing a uniform unit of density.

$$\text{Population density} = \frac{\text{Population of a geographic location}}{\text{Area of that geographic location}}$$

Exercise 22.2
Area of Delhi is 1484 km^2 and population of Delhi as per census 2011 is 16,787,941. Calculate the population density of Delhi.

Solution

$$\text{Population density} = \frac{16,787,941}{1484} = 11,312/\text{km}^2$$

22.3   Measures of Fertility and Reproduction

These are important vital statistics of a population that in some way convey the degree of progress achieved by that population. The measures of fertility and reproduction include:

22.3.1 Crude Birth Rate

Crude birth rate estimates the live births per 1000 or 10,000 population in a given area.

$$\text{Crude Birth Rate} = \frac{\text{Number of live births}}{\text{Population at mid point of time period }(t)} * 1000$$

The crude birth rate does not take into account the age and gender differences in the population.

Exercise 22.3
Midyear population of India in 2011 was 1,210,726,932 and total live birth was 26,108,944. Calculate the birth rate.

Solution

$$\text{Crude Birth Rate} = \frac{26,108,944}{1,210,726,932}$$

$$\text{Crude Birth Rate} = 0.0215$$

$$\text{Crude Birth Rate} = 21.5 \text{ per thousand population}$$

22.3.2 Fertility Rate

22.3.2.1 General Fertility Rate
General fertility rate provides a crude estimate of the fertility rate among females in the reproductive age group, taken as between 15 and 45 years in a population.

$$\text{Genral Fertility Rate (GFR)} = \frac{\text{Number of live births}}{\text{Number of women aged } 15-45 \text{ years}} * 1000$$

22.3.2.2 Age-Specific Fertility Rate
ASFR attempts to refine the estimate of GFR by narrowing the age range of women and, hence, stratifies the fertility rate according to the age groups of eligible females.

$$\text{Age specific fertility rate (ASFR)} = \frac{\text{Number of live births to women aged } t_1 \text{ to } t_2 \text{ years}}{\text{Number of women aged } t_1 \text{ to } t_2 \text{ years}} * 1000$$

22.3.2.3 Total Fertility Rate

Total Fertility Rate = Number of children born to a woman throughout her reproductive age

This gives an estimate of the average number of children that a female can be expected to give birth to during her reproductive phase. TFR can be calculated from ASFR as follows:

Steps to Calculate Total Fertility Rate

1. Calculate the age-specific fertility rates, i.e. number of births per 1000 in a specific age group.
2. Summate the age-specific fertility rates to get $\Sigma ASFR$.
3. Multiply the $\Sigma ASFR$ with the interval of age groups.
4. Divide by 1000 to get the TFR.

22.3.3 Reproduction Rate

These rates provide an estimate of the ability of a population to maintain its reproduction through birth of female children. It is of two types: gross and net reproduction rate.

22.3.3.1 Gross Reproduction Rate

This rate does not take into account the perinatal, neonatal and early childhood death rate of female children before they reach their reproductive phase.

$$\text{Gross Reproduction Rate} = \frac{\text{Number of female children born in the women cohort}}{\text{Number of women in the cohort}}$$

22.3.3.2 Net Reproduction Rate

This rate, in contrast to the gross reproduction rate, adjusts for the perinatal and neonatal mortality of female children.

$$\text{Net Reproduction Rate} = \frac{\text{Number of female children born and survived in the women cohort}}{\text{Number of women survived in the cohort}}$$

22.3.3.3 Sex Ratio at Birth

This ratio is an indicator of the gender equality in a population and in turn affects the reproduction rate.

$$\text{Sex Ratio} = \frac{\text{Number of female live births}}{\text{Number of male live births}} * 1000$$

22.3.3.4 Illegitimacy Rate

$$\text{Illegitimacy rate} = \frac{\text{Number of illegitimate live births}}{\text{Number of unmarried or widowed women aged } 15 - 45 \text{ years}} * 1000$$

Table 22.1 Age-wise data of number of births according to female population

Age	Female population (in 1000)	Number of births	Survival factor
15–19	84.79	2343	0.969
20–24	70.01	14,541	0.967
25–29	72.66	16,736	0.963
30–34	75.92	10,218	0.958
35–39	75.10	5134	0.952
40–44	71.62	1422	0.942
45–49	66.66	93	0.928

Table 22.2 Calculation of age-specific fertility rates

Age	Female population (in 1000)	Number of births	Survival factor	ASFR
15–19	84.79	2343	0.969	27.63
20–24	70.01	14,541	0.967	207.70
25–29	72.66	16,736	0.963	230.33
30–34	75.92	10,218	0.958	134.59
35–39	75.1	5134	0.952	68.36
40–44	71.62	1422	0.942	19.85
45–49	66.66	93	0.928	1.40
Total	516.76	50,487		689.87

Exercise 22.4

The birth in a certain country is classified in Table 22.1 according to the age of mother together with the female population in each age group of the childbearing period.

Calculate (a) general fertility rate, (b) total fertility rate and (c) gross reproduction rate assuming that the sex ratio at birth was 104.5 male births to 100 female births.

Solution (Table 22.2)

(a) General fertility rate $= 50{,}487/516.76 = 97.69$ per thousand.

(b) Total fertility rate $= (689.86*5)/1000 = 3.44$.

(c) Gross reproduction rate $= \left(\frac{50487}{516}\right) * \left(\frac{100}{100+104.5}\right) = 47.77$.

22.4 Measures of Morbidity

For any given disease in a population, the following measures of morbidity due to that particular disease can be calculated.

22.4.1 Disease Incidence Rate

$$\text{Disease Incidence Rate} = \frac{\text{Number of new cases of the disease in time } (t)}{\text{Population in time } (t)} * 1000$$

22.4.2 Disease Prevalence Rate

$$\text{Period Prevalance} = \frac{(\text{Number of new cases} + \text{Number of old cases}) \text{ in time period } (t)}{\text{Population in time period } (t)} * 1000$$

$$\text{Point Prevalance} = \frac{(\text{Number of new cases} + \text{Number of old cases}) \text{ at time point } (t)}{\text{Population at time point } (t)} * 1000$$

Incidence and prevalence are the two most important measures to be calculated for any disease. These indicate the degree of public health importance of a given disease and, hence, are quite helpful to the health authorities in planning of health-care resource allocation.

Exercise 22.5
The population of a town is 25,000 in year 2017. Of these 500 individuals have tuberculosis, and 72 new cases of tuberculosis are diagnosed in same year. Calculate the incidence rate and prevalence rate of tuberculosis in this town.

$$\text{Incidence Rate} = \frac{72}{25,000} * 1000$$

$$\text{Incidence Rate} = 2.88 \text{ per 1000 population}$$

$$\text{Prevalance} = \frac{500 + 72}{25,000} * 1000$$

$$\text{Prevalance} = 22.8 \text{ per 1000 population}$$

22.5   Measures of Mortality

22.5.1 Crude Death Rate

Crude death rate is the proportion of number of deaths (due to any cause) per 1000 population in a given area.

$$\text{Crude Death Rate} = \frac{\text{Total number of deaths in time } (t)}{\text{Population at mid point of time } (t)} * 1000$$

22.5.2 Standardized Death Rate

The weighted average death rates of all the age and sex groups of standard million give the standardized death rate of a place where weights are calculated from standard population. Steps to calculate standardized death rate:

1. Calculate the standard population weights ($W_x = \frac{P_x}{\text{Total population}}$ is the weight at age group x, as calculated from standard population, P_x being population of age group x).
2. Calculate the "age-specific death rate", of the given population Rx.
3. Multiply to get standardized death rate for given age group $StdR_x = R_x * W_x$.
4. Total standardized death rate $= \Sigma StdR_x$.

Exercise 22.6
Calculate the standardized death rate from the following data (Table 22.3).

Solution (Table 22.4)
Crude death rate of given population is $21{,}054/2760 = 7.63$ per thousand population, whereas standardized death rate is 7.44 per thousand population.

22.5.3 Specific Death Rate

Apart from crude death rate, certain specific death rates are of significance in the demographic reports of a region.

Table 22.3 Data of given population and standard population

Age group (years)	Population (in 1000)	Deaths	Standard population (in 1000)
0 to 5	200	1400	98
5 to 10	300	2100	95
10 to 15	350	2400	89
15 to 20	220	1680	87
20 to 25	225	1640	97
25 to 30	200	1390	101
30 to 35	180	1310	115
35 to 40	190	1360	111
40 to 45	200	1489	108
45 to 50	225	1710	120
50 to 55	150	990	125
55 to 60	100	725	130
60 to 65	95	689	150
65 to 70	75	540	125
>70	50	361	100

Table 22.4 Calculation of standardized death rates for various age groups

Age group (in years)	Population (in 1000)	Deaths	Standard population (in 1000)	Weights W_x	Present death rates	$StdR_x$
0 to 5	200.00	2100.00	60.00	0.04	10.50	0.40
5 to 10	300.00	2400.00	90.00	0.06	8.00	0.46
10 to 15	350.00	2400.00	70.00	0.04	6.86	0.30
15 to 20	220.00	1790.00	87.00	0.06	8.14	0.45
20 to 25	225.00	1800.00	82.00	0.05	8.00	0.42
25 to 30	200.00	1390.00	86.00	0.05	6.95	0.38
30 to 35	180.00	1310.00	95.00	0.06	7.28	0.44
35 to 40	190.00	1360.00	90.00	0.06	7.16	0.41
40 to 45	200.00	1489.00	100.00	0.06	7.45	0.47
45 to 50	225.00	1710.00	120.00	0.08	7.60	0.58
50 to 55	150.00	990.00	140.00	0.09	6.60	0.58
55 to 60	100.00	725.00	135.00	0.09	7.25	0.62
60 to 65	95.00	689.00	145.00	0.09	7.25	0.67
65 to 70	75.00	540.00	130.00	0.08	7.20	0.59
> 70	50.00	361.00	150.00	0.09	7.22	0.69
Total	2760.00	21054.00	1580.00	1.0		7.44

22.5.3.1 Age-Specific Death Rate

This estimates the death rate in different age groups in a population.

$$\text{Age Specific Death Rate} = \frac{\text{Number of deaths in specified age group}}{\text{Mid year population of the same age group}} * 1000$$

22.5.3.2 Infant Mortality Rate

Infant mortality rate (IMR) is an important indicator of health-care facilities available to a population.

$$\text{IMR} = \frac{\text{Number of deaths under 1 year of age}}{\text{Number of Live births}} * 1000$$

22.5.3.3 Fetal Death Ratio

$$\text{Fetal Death ratio} = \frac{\text{Number of fetal deaths}}{\text{Number of live births}} * 1000$$

22.5.3.4 Stillbirth Rate or Late Fetal Death Rate

$$\text{Still birth rate or late fetal death rate} = \frac{\text{Number of fetal deaths after 28 weeks of gestation}}{\text{Number of live births + still births}}$$

22.5.3.5 Perinatal Mortality Rate

$$\text{Perinatal Mortality Rate} = \frac{\text{Number of Perinatal deaths}}{\text{Number of live births + still births}} * 1000$$

22.5.3.6 Neonatal Mortality Rate

$$\text{Neonatal Mortality Rate} = \frac{\text{Number of neonatal deaths}}{\text{Number of live births}} * 1000$$

22.5.3.7 Post-neonatal Mortality Rate

$$\text{Post neonatal Mortality Rate} = \frac{\text{Number of deaths between 28 days to 1 year of life}}{\text{Number of live births}} * 1000$$

22.5.3.8 Maternal Mortality Rate

$$\text{Maternal mortality rate} = \frac{\text{Number of deaths due to complication of pregnancy, child birth or puerperium}}{\text{Number of live births}} * 1000$$

22.5.3.9 Sex-Specific Death Rate

$$\text{Sex specific death rate (male)} = \frac{\text{Number of male deaths}}{\text{Mid year population of males}} * 1000$$

22.5.3.10 Age- and Sex-Specific Death Rate

Age and sex specific death rate (male, specific age)

$$= \frac{\text{Number of male deaths in specified age group}}{\text{Mid year population of males in same age group}} * 1000$$

22.5.3.11 Cause-Specific Mortality Rate

$$\text{Cause specific mortality rate} = \frac{\text{Number of death due to a cause}}{\text{Mid year population}} * 1000$$

22.5.3.12 Proportional Mortality Rate

$$\text{Proportional Mortality rate} = \frac{\text{Number of deaths due to a cause}}{\text{Total number of deaths}} * 1000$$

22.5.3.13 Proportional Mortality Indicator (50 Years)

$$\text{Proportional Mortality indicator} = \frac{\text{Number of deaths of persons aged 50 or more}}{\text{Total number of deaths}} * 1000$$

22.5.3.14 Case Fatality Rate

$$\text{Case fatality rate} = \frac{\text{Number of deaths due to disease}}{\text{Number of cases of same disease}} * 1000$$

Exercise 22.7

Study conducted in a town having population of 100,000. The following data was obtained. Calculate MMR, IMR, NMR, postneonatal mortality rate and perinatal mortality rate (Table 22.5).

Solution

$$\text{MMR} = \frac{12}{2700} * 1000 = 4.44$$

$$\text{IMR} = \frac{150}{2700} * 1000 = 55.5$$

$$\text{NMR} = \frac{92}{2700} * 1000 = 34.1$$

Table 22.5 Maternal and child deaths of a particular population

1. Female death due to complications of pregnancy	12
2. Death of children below 1 year	150
3. Death under 28 days	92
4. Death under first 7 days of life	71
5. Death during 7 days to less than 28 days	20
6. Death 28 days to less than 365 days	56
7. Live births	2700
8. Stillbirths	23

$$\text{post neonatal mortality rate} = \frac{56}{2700} * 1000 = 20.7$$

$$\text{perinatal mortality rate} = \frac{23 + 71}{2700 + 23} * 1000 = 34.52$$

Exercise 22.8

The number of new dengue cases is 5510 in a town with 1 Lakh population. Out of these, 75 patients died. Find out the case fatality rate and cause-specific death rate.

Solution

$$\text{Case fatality rate} = \frac{75}{5510} * 1000$$

Case fatality rate = 13.6 per 1000 dengue cases

$$\text{Cause specific death rate} = \frac{75}{100,000} * 1000$$

Cause specific death rate due to dengue = 0.75 per 1000 population

22.6 Measures Related to Population Control

22.6.1 Couple Protection Rate (CPR)

It is the indicator of prevalence of contraceptive practice in the community. It is defined as the percentage of eligible couples protected against childbirth by any one approved methods of family planning. Demographic goal of net reproduction rate (NRR) = 1 can be achieved only if CPR exceeds 60%.

22.6.2 Pearl Index

It is defined as the number of contraceptive failures per 100 woman-years (HWY) of exposure (cohabitation).

$$\text{Pearl Index} = \frac{\text{Total accidental pregnancies irrespective of its outcome}}{\text{Total months of exposure}} * 1200$$

Exercise 22.9

In a study to evaluate effectiveness of a new contraceptive pill done in 250 women over 18-month period, total accidental pregnancies are 25. Calculate the Pearl Index.

Solution

$$\text{Pearl Index} = \frac{25}{250 * 18} * 1200$$

Pearl Index $= 6.66$ per hundred women years

22.7 Measures of Health

22.7.1 Comprehensive Indicators

1. Proportional mortality indicator (with respect to age 50 years).
2. Expectation of life (worked out by the life tables see Chap. 24).
3. Number of females.
4. Crude death rate.

22.7.2 Specific Indicators

Various specific death rates as described above.

Infectious Disease Epidemiology

23

Abstract

Background
Primary attack rate
Secondary attack rate
Case fatality rate
Cause-specific mortality rate
Infection rate

Keywords

Attack rate · Case fatality rate · Specific mortality rate · Infection rate

23.1 Background

Epidemiological data of various infectious diseases is required by the government and policy makers to identify the severity and scale of a given disease in a population. It helps in taking appropriate immediate measures as well as assists in future planning to effectively contain the incidence of a disease. The various terminologies commonly used to understand infectious disease epidemiology are as follows:

(i) *Index case*: Person with a given disease who comes to the attention of public health authorities.

(ii) *Primary case*: Person who acquires the disease from an exposure.

(iii) *Primary attack rate*.

(iv) *Secondary case*: Person who acquires the disease from an exposure to the primary case.

(v) *Secondary attack rate*.

23.2   Primary Attack Rate

Primary attack rate is the measure of morbidity or the speed with which the disease spreads in a population at risk. It is also referred to as incidence proportions, risk and probability of developing disease or cumulative incidence.

Attack rate/incidence rate is usually calculated during a disease outbreak in a narrowly defined population over a short period of time in order to assess the severity of the causative organism.

$$\text{Attack Rate (AR)} = \frac{\text{Number of new cases of disease in population during given time period}}{\text{Number of persons at risk during the given time period}} * 100$$

Exercise 23.1

You are given the numbers of two categories of people who went to a restaurant and either fell ill or were well. Calculate the attack rate for illness after taking a particular food at the restaurant (Table 23.1).

Solution

AR of group who ate the food.

$$AR = \frac{14}{20} * 100.$$

$$AR = 70\%$$

AR of group who did not ate the food

$$AR = \frac{7}{20} * 100$$

$$AR = 35\%$$

Exercise 23.2

During a chikungunya outbreak in a population of 18,000, the number of new cases reported was 500. Calculate the attack rate of chikungunya in this population.

Table 23.1 Data of illness after food intake

Ate the food (exposed)			Did not eat the food (not exposed)		
Ill	Well	Total	Ill	Well	Total
14	6	20	7	13	20

Solution

$$AR = \frac{500}{18000} * 100$$

$$AR = 2.78\%$$

23.3 Secondary Attack Rate (SAR)

It represents the number of exposed individuals developing the disease within the range of incubation period after exposure to the primary case. It is used to estimate the spread of a given disease in a family, household, dormitory or other group environments. It measures the infectivity of the agent and the effects of prophylactic agents like vaccines.

$$SAR = \frac{\text{Total number of cases} - \text{initial case (s)}}{\text{Number of suseptible individuals in the group} - \text{intial case (s)}} * 100$$

Exercise 23.3

In an outbreak of mumps, 20 persons in 15 different households become ill. The population of the community is 1500. One incubation period later, 28 more persons in the same households of primary cases developed mumps. The total number of individuals in these 15 households is 80. Calculate the secondary attack rate.

Solution

$$SAR = \frac{28}{80 - 20} * 100$$

$$SAR = 46.67\%$$

23.4 Case Fatality Rate (CFR)

Case fatality rate represents the proportion of deaths due to a certain disease among the individuals inflicted with that disease. It reflects the lethality (deadliness) of a disease, which can potentially be altered by efficacy of the treatment.

$$CFR = \frac{\text{Number of deaths due to a disease}}{\text{Number of cases of that disease}} * 100$$

23.5   Cause Specific Mortality Rate (CSMR)

It is the mortality rate due to a specified disease/cause for a population. Cause-specific mortality rate for a number of diseases helps to identify the ones that constitute important public health problem and, thus, require main attention of health authorities.

$$CSMR = \frac{\text{Number of deaths due to a disease in a given time period}}{\text{Population during the given time period}} * 100$$

Exercise 23.4
Among a population of 1000 people, 7 out of 30 people having cholera died from the disease in 1 year. Calculate the case fatality rate of cholera and cause-specific mortality rate for cholera.

Solution

$$CFR = \frac{7}{30} * 100$$

$$CFR = 23.3\%$$

$$CSMR = \frac{7}{1000} * 100$$

$$CSMR = 0.7\%$$

23.6   Infection Rate (IR)

It represents the rate of infection due to a certain disease in exposed individuals. This includes both clinically apparent and non-manifested cases. The inapparent infection is generally determined by serological examination.

$$IR = \frac{\text{Manifested cases} + \text{cases with inapparent infection}}{\text{Exposed subjects}} * 1000$$

Life Tables

24

Abstract

Life tables

Life expectancy

Keywords

Life tables · Life expectancy

24.1 Introduction to Life Table

Life tables are the mortality tables depicting the probability of death (mortality probability) of a person at a particular age. Life table shows the probability that a person dies before his/her next birthday.

24.2 What Is Life Expectancy

Life expectancy is the time period (years) an organism is expected to live at a particular age in life.

24.2.1 Uses of Life Tables and Life Expectancy

- Life expectancy is one of the comprehensive indicators of health (see Chap. 22).
- Life expectancy can be used to compare overall mortality of two geographical regions.
- Life expectancy is also used by insurance companies to calculate the *premium* and by actuary departments.

© Springer Nature Singapore Pte Ltd. 2019

S. K. Yadav et al., *Biomedical Statistics*,

https://doi.org/10.1007/978-981-32-9294-9_24

Simplified Example

A bacterial cohort can be easily followed up. Table 24.1 shows artificial data on bacterial cohort comprising of 1000 bacteria followed up for 8 days. The symbols are explained in Table 24.2.

Unlike bacterial cohort described in previous example, human cohorts cannot be practically followed until depletion due to death. Hence, an indirect method is used for making life tables.

At the start, data as shown in Tables 24.3 and 24.4 is utilized for human population.

Table 24.1 Follow up of bacterial cohort

Age	l_x	d_x	q_x	L_x	T_x	e_x
0	1000	490	0.490	755	1619	1.619
1	510	190	0.373	415	864	1.694
2	320	170	0.531	235	449	1.403
3	150	76	0.507	112	214	1.427
4	74	38	0.514	55	102	1.378
5	36	16	0.444	28	47	1.306
6	20	11	0.550	14.5	19	0.950
7	9	9	1.000	4.5	4.5	0.500
8	0	–	–	–	–	–
		$d_x = l_x - l_{x+1}$	$q_x = \frac{d_x}{l_x}$	$L_x = l_{x+1} + 0.5 dx$	$T_x = T_{x+1} + L_x$	$e_x = \frac{T_x}{l_x}$

Table 24.2 Description of symbols used

Symbol	Description of symbol
l_x	Number of organism present at time x
d_x	Number of organism died during time period x
q_x	Mortality probability
L_x	Number of life years lived at time x
T_x	Total number of life years lived
e_x	Life expectancy

Table 24.3 Data of age-wise deaths of a population

Age	Deaths	Population
x	D_x	P_x
0	D_0	P_0
1	D_1	P_1
2	D_2	P_2
:	:	:
99	D_{99}	P_{99}
100 (=100+)	D_{100}	P_{100}

Table 24.4 Initial step in making of life table for human population. The last column (l_x) is sufficient to calculate life expectancy as shown in Table 24.1

Age (in years)	Population	Deaths	M_x	q_x	p_x	l_x
0	457,900	3465	0.0076	0.0075	0.9925	**100,000**
1	534,769	245	0.0005	0.0005	0.9995	99246.14
2	487,657	196	0.0004	0.0004	0.9996	99200.68
3	397,648	121	0.0003	0.0003	0.9997	99160.82
4	329,878	150	0.0005	0.0005	0.9995	99130.65
5	319,873	119	0.0004	0.0004	0.9996	99085.58

From this mortality rate of each group is calculated as

$$M_x = \frac{D_x}{P_x}$$

It is followed by calculation of mortality probabilities as follows

$$q_x = \frac{M_x}{1 + 0.5 M_x} = \frac{D_x}{P_x + 0.5\, D_x}$$

From the mortality probabilities, survival probabilities is calculated as

$$p_x = 1 - q_x$$

After this initial calculation, life table is made by taking "number base" (radix) as 100,000 as starting cohort at 0 age group. Subsequent numbers surviving are calculated as

$$l_x = l_{x-1} * p_{x-1}$$

After this, further calculations are similar to those described for bacterial colonies. To compute the life years (L_x) lived between ages x and x-1

$$L_x = l_{x+1} + a_x d_x$$

$$a_x = 0.5$$

For age 0 $a_x = a_0 = 0.3$ or 0.1.
For last $a_x = a_z = 1/M_z$.

Exercise 24.1
Calculate the life expectancy at birth from data shown in Table 24.5.

Table 24.5 Data of cohort (l_x) and total number of life year lived (T_x)

Age (x)	l_x	T_x
0	1272	?
1	1120	?
2	1080	58,472
3	1050	57,407
4	1042	56,361

Table 24.6 Calculations for life expectancy

Age (x)	l_x	L_x	T_x
0	1272	1196	60,768
1	1120	1100	59,572
2	1080	1065	58,472
3	1050	1046	57,407
4	1042		56,361

Solution

Number of life years lived at time x are calculated in Table 24.6.

$$e_0 = \frac{T_0}{l_0}$$

$$e_0 = \frac{60768}{1272}$$

$$e_0 = 47.8$$

Hence, life expectancy at birth is 47.8 years.

Part IV

Concept of Probability

Introduction to Probability

25

Abstract

Background
Axiomatic approach using set function
Theorems of probability
The addition theorem
The multiplication theorem
Bayes' theorem

Keywords

Theorems of probability · Addition theorem · Multiplication theorem · Bayes' theorem

25.1   Background

The probability of a given event is an expression of the likelihood of occurrence of that event. It is given in numerical form, from 0 to 1. Where 0 means that the event will not occur and 1 means that the event is certain to happen. Probability can never be less than 0 or negative. For random events which are equally likely to occur, Laplace (*classical approach*) definition says "Probability is the ratio of the number of 'favourable' cases to the total number of cases."

$$p(A) = \frac{a}{n} \qquad \text{(Formula 25.1)}$$

where

a = number of ways event A can happen.
n = a total number of equally likely mutually exclusive happenings.

The probability $p(A)$ will be more accurate if the number of trials tends to infinity. However, in practicality, this does not happen. Hence, $p(A)$ always tends to have a degree of inaccuracy built within it. Sometimes, data about trials is not available, and opinion of experts is taken (*subjective approach*). Subjective probability is defined as the probability assigned to an event by an individual based on whatever evidence is available to him. This approach is frequently used by managers.

25.2   Axiomatic Approach Using Set Function

This approach was given by Kolmogorov who introduced probability as set function, saying:

(a) Axiom 1: The probability of an event ranges from 0 to 1.
(b) Axiom 2: The probability of entire sample space is 1, i.e. $p(S) = 1$.
(c) Axiom 3: If A and B are mutually exclusive (or disjoint) events, then the probability of occurrence of either A or B denoted by $p(A \cup B)$ shall be given by

$$p(A \cup B) = p(A) + p(B) \qquad \text{(Formula 25.2)}$$

25.3   Definitions

Given below are some terms that are frequently used while calculation of probabilities for an event.

A. *Experiment:* It refers to the process which results in different possible outcomes or observations.
B. *Event:* It is the outcome of an experiment.
C. *Mutually exclusive events:* Two events are said to be mutually exclusive when both cannot happen simultaneously in a single trial or, in other words, the occurrence of any one of them excludes the occurrence of the other.
D. *Independent events:* Two events are said to be independent when outcome of one does not affect and is not affected by the other.
E. *Dependent events:* Two events are said to be dependent when the occurrence or nonoccurrence of one event in any one trial affects the probability of other events in other trials.
F. *Equally likely events:* When one event does not occur more often than the others.
G. *Simple events:* In case of simple events, we consider the probability of the happening or not happening of single events.
H. *Compound events:* In case of compound events, we consider the joint occurrence of two or more events.

I. *Exhaustive events:* Events are said to be exhaustive when their totality includes all the possible outcomes of a random experiment.

J. *Complimentary events:* Let there be two events A and B. Event A is said to be complimentary event of B if A and B are mutually exclusive and exhaustive, i.e. to say that only one of A or B can occur and that A and B are the only two outcomes of this experiment.

K. *Conditional probability:* If two events A and B are dependent, then the probability of occurrence of A given that B has happened is known as conditional probability.

$$p(A/B) = \frac{p(A \cap B)}{p(B)} \qquad \text{(Formula 25.3)}$$

25.4 Theorems of Probability

Probability of an event in a given experiment can be calculated using one of the three theorems, depending on the situation:

A. The addition theorem.
B. The multiplication theorem.
C. Bayes' theorem.

25.4.1 The Addition Theorem

For mutually exclusive events:

$$p(AorB) = p(A \cup B) = p(A) + p(B) \qquad \text{(Formula 25.4)}$$

For events which are not mutually exclusive:

$$p(AorB) = p(A \cup B) = p(A) + p(B) - P(A \cap B) \qquad \text{(Formula 25.5)}$$

Exercise 25.1
Calculate the probability of drawing a king or a heart from a pack of cards.

Solution

$$p(\text{king}) = \frac{4}{52} \text{ given that there are 4 kings in a pack.}$$

$$p(\text{heart}) = \frac{13}{52} \text{ since there are 13 cards with heart in a pack}$$

$$p(\text{king and heart}) = \frac{1}{52}$$

$$p(\text{king or heart}) = \frac{4}{52} + \frac{13}{52} - \frac{1}{52}$$

25.4.2 The Multiplication Theorem

(a) For independent events:

$$p(A \text{ and } B) = p(A \cap B) = p(A) * p(B) \qquad \text{(Formula 25.6)}$$

(b) Extension of multiplication theorem:

$$p(\text{happening of atleast one of the events})$$
$$= 1 - p \text{ (happening of none of the events)}$$

(c) For dependent events:

$$p(A \text{ and } B) = p(A) * p(B/A) = p(B) * p(A/B) \qquad \text{(Formula 25.7)}$$

Exercise 25.2
A man wants to marry a girl having the following qualities: white complexion ($p = 1/25$) and westernized manners ($p = 1/100$). Find out the probability of him getting married to such a girl when the possession of these attributes is independent.

Solution

$$p \ (A \text{ and } B) = p(A) * p(B) = \frac{1}{25} * \frac{1}{100} = \frac{1}{2500}$$

Exercise 25.3

A problem in statistics is given to five students A, B, C, D and E; their respective probabilities of solving it are 1/2, 1/4, 1/4, 1/6 and 1/7. What is the probability that the problem is solved?

Solution

We need to calculate: p (happening of at least one of the events) $= 1 - p$ (happening of none of the events)

p (A fails) $= 1 - p$(A solves) $= 1 - 1/2 = 0.5$
p (B fails) $= 1 - p$(B solves) $= 1 - 1/4 = 0.75$
p (C fails) $= 1 - p$(C solves) $= 1 - 1/4 = 0.75$
p (D fails) $= 1 - p$(D solves) $= 1 - 1/6 = 0.83$
p (E fails) $= 1 - p$(E solves) $= 1 - 1/7 = 0.86$
p (All fails) $= p$ (A fails and B fails and C fails and D fails and E fails) $= p$ (A fails) $*$
    p (B fails) $* p$ (C fails) $* p$ (D fails) $* p$ (E fails) $= 0.2$
p (Problem solved by at least one) $= 1 - p$ (All fails) $= 1 - 0.2 = 0.8$

Exercise 25.4

A bag contains six white and three black balls. Two balls are drawn at random one after the other without replacement. Find the probability that both balls drawn are black.

Solution

We have to find $p(A$ and $B)$

p of black ball in first attempt:

$$p(A) = \frac{3}{9}$$

p of black ball in second attempt:

$$p(B/A) = \frac{2}{8}$$

Hence,

$$p(A \text{ and } B) = p(A) * p(B/A)$$

$$p(A \text{ and } B) = \frac{6}{72} = \frac{1}{12} = 0.083$$

25.4.3 Bayes' Theorem (Fig. 25.1)

$$p\left(^{A_1}\!\!/_B\right) = \frac{p(A_1 \text{ and } B)}{p(B)}$$

$$p\left(^{A_2}\!\!/_B\right) = \frac{p(A_2 \text{ and } B)}{p(B)}$$

From the figure it is clear that

$$p(B) = p(A_1 \text{ and } B) + p\,(A_2 \text{ and } B)$$

Also, we know that for dependent events, rule of multiplication states that

$$p(A_1 \text{ and } B) = p\,(A_1) * p\left(^{B}\!\!/_{A_1}\right)$$

$$p(A_2 \text{ and } B) = p(A_2) * p\left(^{B}\!\!/_{A_2}\right)$$

and also

$$p(B) = \left\{ p(A_1) * p\left(\frac{B}{A_1}\right) \right\} + \left\{ p(A2) * p\left(\frac{B}{A_2}\right) \right\}$$

Hence:

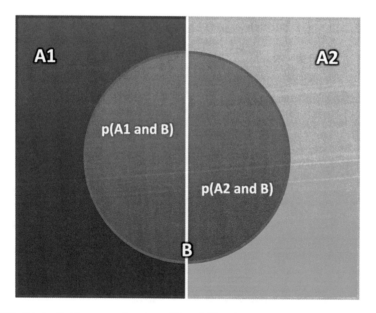

Fig. 25.1 Schematic diagram explanation of Bayes' Theorem

$$p\left(^{A_i}\!/_B\right) = \frac{\left[p(A_i) * p\left(^B\!/_{A_i}\right)\right]}{p(B)} \qquad \text{(Formula 25.8)}$$

$$p\left(^{A_i}\!/_B\right) = \frac{\left[p(A_i) * p\left(^B\!/_{A_i}\right)\right]}{\left\{\Sigma_i\left[p(A_i) * p\left(^B\!/_{A_i}\right)\right]\right\}} \qquad \text{(Formula 25.9)}$$

Exercise 25.5

A manufacturing firm produces units of a product in four plants. Define event A_i, unit is produced in plant i, $i = 1,2,3,4$, and event B, unit is defective.

From the past records the proportions of defective units produced at each plant are as follows: $p(B/A_1) = 0.02$; $p(B/A_2) = 0.10$; $p(B/A_3) = 0.25$; $p(B/A_4) = 0.03$.

The proportions of total units produced by plants are as follows: $A_1 = 24\%$; $A_2 = 16\%$; $A_3 = 35\%$; $A_4 = 25\%$.

If a unit tested was found to be defective, what is the probability that the unit was produced in plant 3?

Solution

(Table 25.1) Bayes' theorem is

$$p\left(^{A_i}\!/_B\right) = \frac{\left[p(A_i) * p\left(^B\!/_{A_i}\right)\right]}{\left\{\Sigma_i\left[p(A_i) * p\left(^B\!/_{A_i}\right)\right]\right\}}$$

We need to find $p(A_3/B)$, which is,

$$p\left(^{A_3}\!/_B\right) = \frac{\left[p(A_3) * p\left(^B\!/_{A_3}\right)\right]}{\left\{\Sigma_i\left[p(A_i) * p\left(^B\!/_{A_i}\right)\right]\right\}}$$

$$p\left(^{A_3}\!/_B\right) = 0.76$$

Table 25.1 Calculation of conditional probabilities

Plant (i)	$p(A_i)$	$p(B/A_i)$	$p(A_i)*p(B/A_i)$	$p(A_i)*p(B/A_i)/p(B)$
1	0.24	0.02	0.00	0.04
2	0.16	0.1	0.02	0.14
3	0.35	0.25	0.09	**0.76**
4	0.25	0.03	0.01	0.06
Total	1	–	0.12	1.00

Random Variable and Mathematical Expectation

26

Abstract

Random variable
Mathematical expectation

Keywords

Random variable · Mathematical expectation

26.1 Random Variable

A variable whose value is determined by the outcome of a random experiment is called a random variable, a chance variable or a stochastic variable. A random variable may be discrete or continuous.

If a variable x can assume a discrete set of values $x_1, x_2, \ldots x_k$ with respective probabilities $p_1, p_2, \ldots p_k$, where $\sum_k p_k = 1$, then function $p(X)$ is called probability density function (PDF) or frequency function of x.

$$PDF = p(X)$$

Cumulative density function:

$$CDF(t \leq x) = \int_{t=-\infty}^{x} PDF(t)dt \text{ for continuous distribution.}$$

$$CDF(t \leq x) = \sum_{t=-\infty}^{x} PDF(t) \text{ for discrete distribution.}$$

26.2   Mathematical Expectation

The mathematical expectation (also called the expected value) of a random variable is the weighted arithmetic mean of the variable; the weights are all the respective probabilities of the values that the variable can possibly assume. Where p_k is the probability of the *kth* value (x_k)

$$E(x) = \Sigma_k p_k x_k$$

Exercise 26.1

The probability that a man fishing at a particular place will catch 1, 2, 3 and 4 fishes is 0.45, 0.25, 0.2 and 0.1, respectively. What is the expected number of fishes caught?

Solution

$$E(x) = 0.45 * 1 + 0.25 * 2 + 0.2 * 3 + 0.1 * 4 = 1.65.$$

Statistical Distribution: Discrete

27

Abstract

Binominal Distribution

Keyword

Binominal distribution

27.1 Binominal Distribution

It is also known as Bernoulli distribution. The following assumptions are made:

1. The number of trials will be n.
2. The outcome is dichotomous and mutually exclusive.
3. The probability of success in a single trial is p.
4. The probability of failure in a single trial, $q = 1-p$.
5. The outcomes/trials are independent of each other.

Then, the probability of r success in n trials is given by

$$1. \quad p(r) = {}^nC_r p^r q^{(n-r)} \qquad \text{(Formula 27.1)}$$

where

$$ {}^nC_r = \frac{n!}{[r!(n-r)!]} $$

- For $r = 0$, $p(0) = q^n$.
- For $r = n$, $p(n) = p^n$

$$2. \quad \text{CDF } (r \leq x) = \sum_{r=0}^{x} \text{PDF}(r)$$

27.1.1 Properties of Binomial Distribution

$$\text{Mean} = np \qquad \qquad \text{(Formula 27.2)}$$

$$SD = \sqrt{npq} \qquad \qquad \text{(Formula 27.3)}$$

If n is large and if neither p nor q is too close to zero, the binomial distribution can be closely approximated to a normal distribution (Fig. 27.1) with standard normal deviate as

$$z = \frac{(X - np)}{\sqrt{npq}} \qquad \qquad \text{(Formula 27.4)}$$

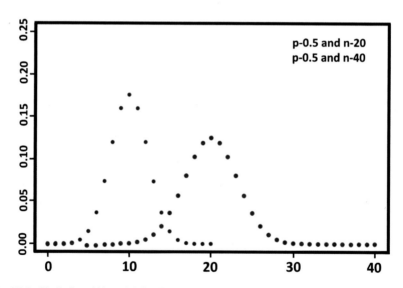

Fig. 27.1 Similarity of binomial distribution to normal distribution for larger n

Shape of the Binomial Distribution

When $p = 0.5$, the binomial distribution is symmetrical – the mean and median are equal.

Even if $p \neq 0.5$, the shape of the distribution tends to be symmetrical with larger value of N.

Exercise 27.1

A dice is thrown 6 times. Considering that getting an odd number is a success, what is the probability of five successes?

Solution

$$n = 6$$

$$r = 5$$

$$p = \frac{3}{6} = \frac{1}{2}$$

$$q = 1 - \frac{1}{2} = \frac{1}{2}$$

$$p(r) = {}^nC_r p^r q^{(n-r)} = {}^6C_5 p^5 q^{(1)} = \frac{3}{32}$$

Univariate Logistic Regression: Theoretical Aspects

28

Abstract

Logistic Regression Model
Estimation of coefficient (data fitting)
Assessing the significance of the model

Keywords

Logit · Logistic regression

28.1   Model

Logistic regression is a special type of generalised linear modelling where the outcome (dependent variable) is binary, i.e. there are two possibilities of the outcome – the event occurs or does not occur. As we have seen in the linear regression model (see Chap. 16), the outcome variable is related to independent variables by following function.

$$\beta_0 + \beta_1 x_1$$

However, when outcome is binary, it cannot be directly related to the above stated linear function.

If p is probability that $y = 1$, then, $1 - p$ is probability that $y = 0$.
odds ratio of y $= 1$ is given as

$$\text{odds}\,(y = 1) = \frac{p(y = 1)}{p\,(y \neq 1)} = \frac{p}{1 - p}$$

This odds ratio will vary from 0 to ∞. The log of odds ratio is given as

© Springer Nature Singapore Pte Ltd. 2019
S. K. Yadav et al., *Biomedical Statistics*,
https://doi.org/10.1007/978-981-32-9294-9_28

$$\log{}_e(\text{odds}(y = 1)) = \log\left(\frac{p}{1-p}\right)$$

This is known as Logit (p). Logit (p) will vary between $-\infty$ and ∞. Hence, suitable model for logistic regression is as follows:

$$\log{}_e\left(\frac{p}{1-p}\right) = \beta_0 + \beta_1 x_1$$

Probability of y is calculated from the model by (Fig. 28.1)

$$p(y) = \frac{exp(\beta_0 + \beta_1 x_1)}{1 + exp(\beta_0 + \beta_1 x_1)} = \frac{1}{1 + exp^{-(\beta_0 + \beta_1 x_1)}}$$

β_0 is the log odds in favour of $y = 1$.
β_1 is the additive change in log odds for $y = 1$ when x_1 increases by one unit.

Fig. 28.1 Illustration of logistic regression (blue curve) and linear regression (red line)

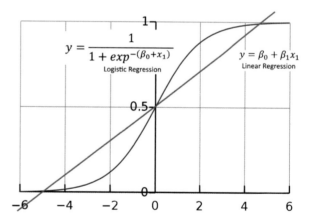

28.2   Estimation of Coefficient (Data Fitting)

For data with "n" observation, the likelihood function is defined as

$$l(\beta) = \prod_{i=1}^{n} p(x_i, \beta)^{y_i} (1 - p(x_i, \beta))^{1-y_i}$$

We maximise the likelihood function to get the parameters.

$$ll(\beta) = \Sigma[y_i \log_e p(x_i, \beta) + (i - y_i) \log_e (1 - p(x_i, \beta))]$$

where

$$\beta = \begin{pmatrix} \beta_0 \\ \beta_1 \end{pmatrix}$$

$$x = \begin{pmatrix} 1 \\ x_1 \end{pmatrix}$$

$$p(x_i, \beta) = \frac{1}{1 + e^{-\beta^T x}}$$

Likelihood formula

$$ll(\beta) = \Sigma \left[y_i \beta^T x_i - \log_e \left(1 + e^{\beta^T x_i} \right) \right]$$

To maximise $ll(\beta.)$

Set $\frac{dll(\beta)}{d\beta_j} = 0$, for each independent variable j starting from β_0 intercept.

$$\frac{dll(\beta)}{d\beta_j} = \Sigma_i [y_i - p(x_i, \beta)] x_{ij} = 0$$

This will result in a set of $p + 1$ non-linear equation. These equations are solved using Newton-Raphson algorithm.

28.3   Assessing the Significance of the Model

1. Wald statistics (W) for significance of independent variables.

$\text{Wald}^2 = \left[\frac{\beta}{SE\beta}\right]^2$ follows chi-square distribution, $df = 1$.

$$W \sim N(0, 1)$$

$$W^2 \sim N^2(1)$$

2. Partial R for significance of independent variables

$$\text{Patial } R = \sqrt{\frac{\text{Wald} - 2}{-2LL(\alpha)}}$$

   Valid for Wald > 2

3. Model chi-square by likelihood ratio test

$$LR_i = -2[LL(\alpha) - LL(\alpha, \beta)]$$

   follows chi-squared distribution with degree of freedom equal to number of independent variables.

4. Model pseudo R^2

$$\text{McFadden's } R^2 = 1 - \left[\frac{LL(\alpha, \beta)}{LL(\alpha)}\right]$$

R^2 varies from 0 to1.

Part V

Computers in Statistics

Use of Computer Software for Basic Statistics

29

Abstract

Installing R and RStudio
General function in R
Data types
Data classification
Data presentation
Measures of central tendency
Measures of location
Measures of dispersion
Statistical distributions – continuous
Test for inference – one sample or two sample mean
Test for inference – multiple sample comparisons
Test for inference – categorical data I
Test for inference – categorical data II
Test for inference – correlation and regression
Analysis of diagnostic tests
Logistic regression
Non parametric tests

Keywords

R programming · RStudio

29.1 Installing R and RStudio

In this chapter we shall briefly describe the use of a free-to-download software available for statistics.

R is a very useful open-source program meant for statistics as well as graphics. R can be freely downloaded from Cran web site (URL 'https://www.r-project.org'. Last accessed on September 2019). Install R on your computer with default options.

© Springer Nature Singapore Pte Ltd. 2019
S. K. Yadav et al., *Biomedical Statistics*,
https://doi.org/10.1007/978-981-32-9294-9_29

RStudio is an integrated development environment (IDE) for R language. It has a scripting window, a console, and a graphics (plot) window. RStudio can be downloaded from its web site (URL 'https://www.rstudio.com' and installed with default settings. Last accessed on September 2019).

29.1.1 Installing Required Packages

Open RStudio. Go to the "Packages" tab, and click on "Install Packages" to install following packages:

1. e1071
2. dplyr
3. caret

29.2   General Functions in R

In the following demonstration, the data file, namely, "psoriasis.csv" is used, the data of which is reproduced in Appendix 6.

The file contains the data of 42 psoriasis patients. The variables of data used in this chapter are tabulated in Table 29.1.

The results of programming include certain additional outputs, the description of which is beyond the scope of this basic book on statistics. Boxes with solid borders

Table 29.1 Variables used in this chapter from data file psoriasis.csv

Variable	Description	R-syntax[a]
AGE	Age of the patients	mydata$AGE
SEX	Gender of patients	mydata$SEX
SMOKER	Smoking habit of patients	mydata$SMOKER
ALCOHOLIC	History of alcohol intake	mydata$ALCOHOLIC
STRESS	Stress level of patients	mydata$STRESS
LOCATION	Location of lesions	mydata$LOCATION
PASI	Psoriasis area severity index of patients	mydata$PASI
WBC	White blood cell count	mydata$WBC

[a]mydata is the name given to data frame in the examples in this chapter

represent syntax to be typed in the scripting windows of RStudio. It is noteworthy that R language is case-sensitive. Boxes with dashed border represent output in console of RStudio.

Setting and getting the working directory in which our data file resides

```
setwd("C:\Users\Username\Documents\BasicStatisticsInR")
getwd()
```

Restricting our results to 4 digits after decimal point

```
#Options for post decimal digits
options(digits = 4)
```

Opening the csv data file in R. The file must be in working directory else full path need to be specified. For online data file URL needs to be specified.

```
#Open Data File and View data
mydata <- read.csv('psoriasis.csv')
View(mydata)

#Summary
summary(mydata)
```

Summary will be as follows:

```
 AGE              SEX      SMOKER          ALCOHOLIC        STRESS
 Min.    :13.0    F:15    Min.    :0.000   Min.    :0.000   Min.    :1.00
 1st Qu.:22.2    M:27    1st Qu.:0.000   1st Qu.:0.000   1st Qu.:2.00
 Median :31.5            Median :0.000   Median :0.000   Median :3.00
 Mean    :36.1            Mean    :0.333   Mean    :0.238   Mean    :2.71
 3rd Qu.:46.8            3rd Qu.:1.000   3rd Qu.:0.000   3rd Qu.:3.00
 Max.    :90.0            Max.    :1.000   Max.    :1.000   Max.    :4.00
```

29.3   Use of RStudio in Basic Statistical Calculations

29.3.1 Data Types (Also see Chap. 3)

In R, qualitative data is named as "factor", and quantitative data is named as "numeric" (continuous data) or "integer" (discrete data).

```
#Data Type
class(mydata)
class(mydata$AGE)
class(mydata$SEX)
class(mydata$LOCATION)
class(mydata$WBC)
```

Output will be as follows:

```
> #Data Type
> class(mydata)
[1] "data.frame"
> class(mydata$AGE)
[1] "integer"
> class(mydata$SEX)
[1] "factor"
> class(mydata$LOCATION)
[1] "factor"
> class(mydata$WBC)
[1] "numeric"
```

In R, if scores are of "integer" class, they can be converted to factors as follows:

```
#Change the score to 'factor'
class(mydata$STRESS)
mydata$STRESS = factor(mydata$STRESS, levels = c('1','2','3', '4'),
labels = c(1,2,3,4))
class(mydata$STRESS)
```

Output will be as follows:

```
> class(mydata$STRESS)
[1] "integer"
> mydata$STRESS = factor(mydata$STRESS, levels = c('1','2','3', '4'),
labels = c(1,2,3,4))
> class(mydata$STRESS)
[1] "factor"
```

29.3.2 Data Classification (Also see Chap. 4)

29.3.2.1 Frequency Table of Discrete Data

Objective: To obtain frequency of each observation of variable AGE

```
#FrequencyTable of Quantitative-Discrete data
Tbl <- table(mydata$AGE)
#Tbl

cbind(Tbl)
```

Output will be as follows, first column is xi and second column is fi

```
        Tbl
13      3
16      1
18      3
21      3
22      1
23      1
25      3
26      2
27      2
28      1
30      1
33      2
35      1
38      1
40      3
45      2
46      1
47      1
48      1
50      2
57      1
60      1
64      1
65      2
70      1
90      1
```

29.3.2.2 Frequency Table of Continuous Data

Objective: To obtain frequency distribution of continuous variable WBC

```
mywbc = mydata$WBC
range(mywbc) # required to make class intervals

breaks = seq(2, 22, by=2)      # class interval = 2
#breaks

mywbc = cut(mywbc, breaks, right=FALSE)

mywbc.freq = table(mywbc)
#mywbc.freq

cbind(mywbc.freq)
```

Output will be as follows:

```
       mywbc.freq
[2,4)          2
[4,6)          8
[6,8)         15
[8,10)         6
[10,12)        9
[12,14)        1
[14,16)        0
[16,18)        0
[18,20)        0
[20,22)        1
```

29.3.2.3 Frequency Distribution of Categorical Data

Objective: To obtain frequency of male and females in variable SEX

```
#FrequencyTable of Qualitative data
Tbl <- table(mydata$SEX)
#Tbl
dimnames(Tbl) <- list(gender = c("F", "M"))
Tbl
```

Output will be as follows:

```
gender
 F  M
15 27
```

29.3.2.4 Proportions of Categorical Data

Objective: To obtain proportions of male and female in variable SEX

```
#Showing proportions
Tbl <- prop.table(Tbl)
#Tbl
dimnames(Tbl) <- list(gender = c("F", "M"))
Tbl
```

Output will be as follows:

```
gender
     F      M
0.3571 0.6429
```

29.3.2.5 Cross Tabulation

Objective: Cross tabulation of frequencies of SEX and SMOKER

```
#Cross tabulation
Tbl <- table(mydata$SEX, mydata$SMOKER)
#Tbl
dimnames(Tbl) <- list(gender = c("F", "M"), smoker = c("N","Y"))
Tbl
```

Output will be as follows:

```
       smoker
gender  N   Y
     F 15   0
     M 13  14
```

29.3.3 Data Presentation (Also see Chap. 5)

29.3.3.1 Pie Diagram
Objective: To make pie diagram depicting scores of stress in our study

```
#Pie diagram
Tbl <- table(mydata$STRESS)
Tbl
pie(Tbl, labels = c(1,2,3,4), radius = 1, main = "Stress",
border="blue", clockwise = T)
```

Output visible in "plot" will be as follows (Fig. 29.1):

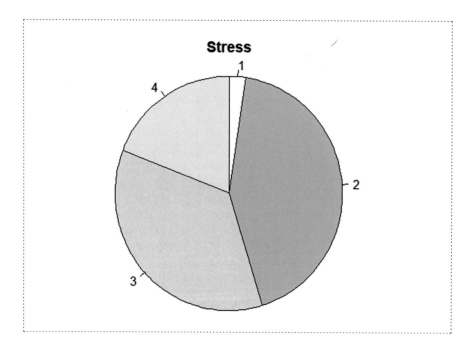

Fig. 29.1 Pie diagram representing score of stress in the psoriasis subjects in our study

29.3.3.2 Bar Diagram

Objective: To make bar diagram depicting scores of stress in our study

```
#Bar Diagram
barplot(table(mydata$STRESS), cex.names = 1.20, main = "Stress")
barplot(prop.table(table(mydata$STRESS)), cex.names = 1.20, main =
"Stress",xlab='scores')
```

Output will be as follows (Fig. 29.2):

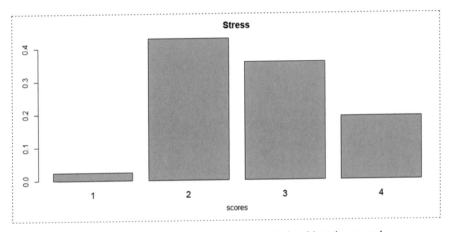

Fig. 29.2 Bar diagram representing scores of stress in psoriasis subjects in our study

29.3.3.3 Histogram and Frequency Polygon

Objective: To make histogram and frequency polygon of variable AGE

```
#Histogram
#hist(mydata$AGE, freq = TRUE, main = "Age")
h = hist(mydata$AGE, xlim = c(0, max(mydata$AGE) + 10), col =
"steelblue3", right = F, main = "Age", xlab='AGE (Years)')
mp = c(min(h$mids) - (h$mids[2] - h$mids[1]), h$mids, max(h$mids) +
(h$mids[2] - h$mids[1]))
freq = c(0, h$counts, 0)
lines(mp, freq, type = "b", pch = 20, col = "red", lwd = 3)
```

Output will be as follows (Fig. 29.3):

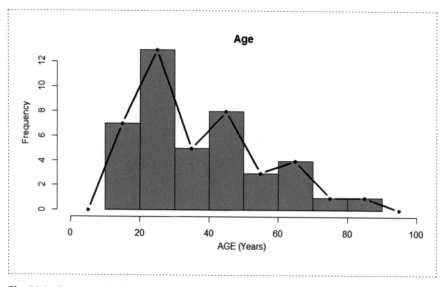

Fig. 29.3 Frequency distribution of variable AGE in our study

29.3.3.4 O'give Curve
Objective: To make less than O'give for variable AGE

```
#O'give curve
h = hist(mydata$AGE, xlim = c(0, max(mydata$AGE) + 10), col =
"Steelblue3", right = F, main = "Age")
ucl = seq(from = min(h$breaks), to = max(h$breaks), by = h$breaks[2] -
h$breaks[1])
ucl = c(0, ucl[-1])
cf = c(0, cumsum(h$counts))
par(bg = "white")
plot(ucl, cf, type = "b", col = "blue", pch = 20,main='Less than
Ogive',xlab='AGE (Years)', ylab = 'Cummulative Frequency')
```

Output will be as follows (Fig. 29.4):

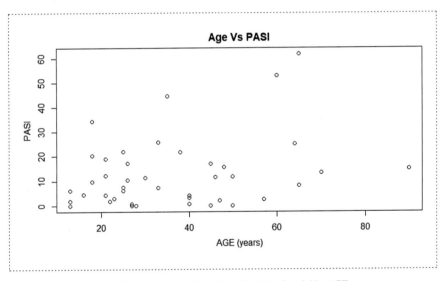

Fig. 29.4 Cummulative frequency graph (less than O'give) of variable AGE

29.3.3.5 Scatter Diagram

Objective: To visualise the correlation between AGE and PASI (if any)

```
#Scatter diagram
plot(mydata$AGE, mydata$PASI, main = "Age Vs PASI", xlab = 'AGE
(years)', ylab = 'PASI')
```

Output will be as follows (Fig. 29.5):

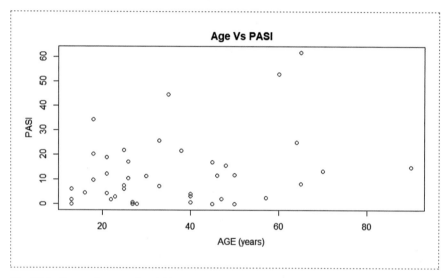

Fig. 29.5 Scatter diagram of variable PASI and AGE to assess correlation (if any)

29.3.3.6 Box and Whisker Plot

Objective: To study the distribution variable AGE

```
#Box Plot
boxplot(mydata$AGE, main = "Age", xlab = 'AGE (years)')
```

Output will be as follows (Fig. 29.6):

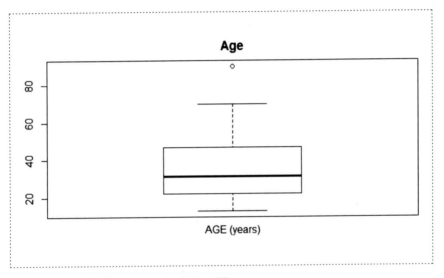

Fig. 29.6 Box and Whisker plot of variable AGE

29.3.4 Measures of Central Tendency (Also see Chap. 6)

29.3.4.1 Mean, Median and Mode

Objective: To measure the mean, median and mode of variable AGE. Also find the mean age of SMOKERS

```
#Mean Median Mode and mean age of smokers
mean(mydata$AGE)
median(mydata$AGE)
mode(mydata$AGE)
mean(mydata$AGE[mydata$SMOKER == '1'])
```

Output will be as follows:

```
> mean(mydata$AGE)
[1] 36.12
> median(mydata$AGE)
[1] 31.5
> mode(mydata$AGE)
[1] "numeric"
> mean(mydata$AGE[mydata$SMOKER == '1'])
[1] 38.57
```

Mode returned "numeric" error since our data is not unimodal.

29.3.5 Measures of Location (Also see Chap. 7)

Objective: To find the first value, 25th percentile (or 2.5th decile or 1st quartile) and 80th percentile (or 8th decile) of variable AGE

In R location is returned with reference to 100 divisions. Data should be first sorted.

```
#Measures of Location
mysort <- sort(mydata$AGE)
quantile(mysort, probs = c(0, 0.25, 0.80))
```

Output is as follows:

```
> #Measures of Location
> mysort <- sort(mydata$AGE)
> quantile(mysort, probs = c(0, 0.25, 0.80))
    0%    25%    80%
 13.00  22.25  49.60
```

29.3.6 Measures of Dispersion (Also see Chap. 8)

29.3.6.1 Dispersion

Objective: To find the range, variance, standard deviation, interquartile range and mean absolute deviation of the variable AGE

```
#Measures of dispersion
#Range
range(mydata$AGE)
#Variance
var(mydata$AGE)
#Standard Deviation
sqrt(var(mydata$AGE))
sd(mydata$AGE)
#Interquartile range
IQR(mydata$AGE)
#Mean absolute deviation
mad(mydata$AGE)
```

Output is as follows:

```
> #Measures of dispersion
> #Range
> range(mydata$AGE)
[1] 13 90
> #Variance
> var(mydata$AGE)
[1] 326.1
> #Standard Deviation
> sqrt(var(mydata$AGE))
[1] 18.06
> sd(mydata$AGE)
[1] 18.06
> #Interquartile range
> IQR(mydata$AGE)
[1] 24.5
> #Mean absolute deviation
> mad(mydata$AGE)
[1] 17.79
```

29.3.6.2 Skewness and Kurtosis (Requires Package "e1071" to be installed)

Objective: To find the skewness and kurtosis of variable AGE

```
#Skewness and kurtosis
library(e1071)
e1071::skewness(mydata$AGE)
e1071::kurtosis(mydata$AGE)

#Outliers
boxplot.stats(mydata$AGE)$out
```

Output will be as follows:

```
> #Skewness and kurtosis
> library(e1071)
> e1071::skewness(mydata$AGE)
[1] 0.8562
> e1071::kurtosis(mydata$AGE)
[1] 0.1454
> #Outliers
> boxplot.stats(mydata$AGE)$out
[1] 90
```

29.3.7 Statistical Distributions: Continuous (Also see Chap. 10)

29.3.7.1 Calculating Z-Score

Objective: To calculate the z-score for AGE = 42 from the data

```
#z-Score
hist(mydata$AGE, main = 'Age') #histogram

#POPULATION PARAMETER CALCULATIONS
pop_sd <- sd(mydata$AGE)*sqrt((length(mydata$AGE)-
1)/(length(mydata$AGE)))
pop_mean <- mean(mydata$AGE)

z <- (42 - pop_mean) / pop_sd
z
```

Output is as follows:

```
> z
[1] 0.3296
```

29.3.7.2 Standardising a Variable

Objective: To standardise the variable AGE.

```
#Standardizing variables
stdAge <- scale(mydata$AGE)
```

Output will be as follows:

```
> #Standardizing variables
> stdAge <- scale(mydata$AGE)
> mean(stdAge)
[1] -8.093e-17
> sd(stdAge)
[1] 1
```

29.3.7.3 Statistical Distributions

For normal distribution probability density function (PDF) can get R using the function dnorm(), whereas cumulative density function (CDF) can be accessed using pnorm(). The inverse of CDF can be acquired using the function qnorm(). Corresponding functions include dt(), pt() and qt() for t-distribution; dchisq(), pchisq () and qchisq() for chi-square distribution; and df(), pf() and qf() for F-distribution.

If z-score is supplied to the function pnorm(), it will return the probability value from $-\infty$ to z.

$$CDF(z) = p(X \leq z) = pnorm(z)$$

$$p(X > z) = 1 - pnorm(z)$$

If probability value from $-\infty$ to z is supplied to the function qnorm(), it will return the corresponding value of z-score.

$$qnorm(p(X \leq z)) = z$$

Objective: To find the z, t, χ^2, F value for 95% area under the curve ($p \leq x$)

```
#Normal distribution, t distribution, chi-square distribution, f
distribution
qnorm(0.95)
qt(0.95, df=30)
qchisq(0.95, df=1)
qf(0.95, df1=10, df2=10)
```

Output will be as follows:

```
> #Normal distribution, t distribution, chi-square distribution, f
distribution
> qnorm(0.95)
[1] 1.645
> qt(0.95, df=30)
[1] 1.697
> qchisq(0.95, df=1)
[1] 3.841
> qf(0.95, df1=10, df2=10)
[1] 2.978
```

Note for two-tailed values of z, t use 0.975 instead of 0.95.

29.3.7.4 Tests for Normality

Objective: To assess whether the variable AGE is normally distributed

```
#Kolmogorov-Smirnov test
with(mydata, ks.test(AGE, "punif"))

#Shapiro-Wilk normality test
with(mydata, shapiro.test(AGE))
```

Output will be as follows:

```
#Kolmogorov-Smirnov test
> with(mydata, ks.test(AGE, "punif"))

        One-sample Kolmogorov-Smirnov test

data:   AGE
D = 1, p-value <2e-16
alternative hypothesis: two-sided

>
> #Shapiro-Wilk normality test
> with(mydata, shapiro.test(AGE))

        Shapiro-Wilk normality test

data:   AGE
W = 0.92, p-value = 0.008
```

Inference: Variable AGE is not normally distributed in the given data.

29.3.8 Test for Inference: One-Sample or Two-Sample Mean (Also see Chap. 12)

29.3.8.1 One-Sample t-Test

Objective: To find if mean of AGE is significantly different from population mean of 30

```
#One sample t test
popmu <- 30 #Population mean
compy <- rep.int(popmu, length(mydata$AGE))
t.test(mydata$AGE, y = compy, alternative = "two.sided", mu = 0,
paired = FALSE, var.equal = FALSE, conf.level = 0.95)
```

Output will be as follows:

```
        Welch Two Sample t-test

data:  mydata$AGE and compy
t = 2.2, df = 41, p-value = 0.03
alternative hypothesis: true difference in means is not equal to 0
95 percent confidence interval:
   0.4921 11.7460
sample estimates:
mean of x mean of y
    36.12     30.00
```

Inference: Mean of AGE is significantly different from population mean of 30 ($p = 0.03$).

29.3.8.2 Two-Sample t-Test

Objective: To assess whether AGE of smokers and non-smokers is significantly different

```
#Two sample t test
temp <- with(mydata, t.test(AGE[mydata$SMOKER == '0'],
AGE[mydata$SMOKER == '1']), alternative = "two.sided", mu = 0, paired
= FALSE, var.equal = FALSE, conf.level = 0.95)
temp
```

Output will be as follows:

```
        Welch Two Sample t-test

data:  AGE[mydata$SMOKER == "0"] and AGE[mydata$SMOKER == "1"]
t = -0.68, df = 34, p-value = 0.5
alternative hypothesis: true difference in means is not equal to 0
95 percent confidence interval:
 -14.604   7.247
sample estimates:
mean of x mean of y
   34.89     38.57
```

Inference: AGE of smokers and non-smokers is not significantly different ($p = 0.5$).

29.3.8.3 Paired t-Test

Objective: To assess whether the following paired observations are significantly different.

Before	21	34	67	54	89	37	12
After	19	30	65	49	79	35	9

```
# #Paired t-test
x <- c(21,34,67,54,89,37,12)
y <- c(19,30,65,49,79,35,9)
t.test(x, y, paired = TRUE, alternative = "two.sided")
```

```
Output will be as follows
Paired t-test

data:  x and y
t = 3.7, df = 6, p-value = 0.01
alternative hypothesis: true difference in means is not equal to 0
95 percent confidence interval:
 1.33 6.67
sample estimates:
mean of the differences
            4
```

Inference: The paired data is significantly different. (p = 0.01).

29.3.9 Test for Inference: Multiple Sample Comparisons (Also see Chap. 13)

29.3.9.1 ANOVA One Way

Objective: To assess whether PASI score is significantly different among various STRESS scores

```
#Anova one way, PASI with STRESS
myanova <- aov(PASI~STRESS, data=mydata)
summary(myanova)
```

Output will be as follows:

```
            Df Sum Sq Mean Sq F value Pr(>F)
STRESS       1    809     809    4.31  0.044 *
Residuals   40   7503     188
---
Signif. codes:  0 '***' 0.001 '**' 0.01 '*' 0.05 '.' 0.1 ' ' 1
```

Inference: PASI score is significantly different among various STRESS scores.

29.3.9.2 ANOVA Two Way

Objective: To assess whether length of tooth *(len)* is significantly different across supplements *(supp)* and across dose *(dose)* using inbuilt data *ToothGrowth*

```
# Store the data in the variable my_data
my_data <- ToothGrowth


res.aov2 <- aov(len ~ supp + dose, data = my_data)
summary(res.aov2)
```

Output will be as follows:

```
> # Store the data in the variable my_data
> my_data <- ToothGrowth
> res.aov2 <- aov(len ~ supp + dose, data = my_data)
> summary(res.aov2)
            Df Sum Sq Mean Sq F value    Pr(>F)
supp         1  205.4   205.4   11.45    0.0013 **
dose         1 2224.3  2224.3  123.99  6.31e-16 ***
Residuals   57 1022.6    17.9
---
Signif. codes:  0 '***' 0.001 '**' 0.01 '*' 0.05 '.' 0.1 ' ' 1
```

Inference: Length of tooth growth is significantly different across supplements ($p = 0.0013$) as well as across dose ($p < 0.000$).

29.3.10 Test for Inference: Categorical Data I (Also see Chap. 14)

29.3.10.1 Tests for Proportions: One Sample

Objective: If the proportion of present study is $\frac{25}{65}$, then assess whether it is significantly different from a population proportion of 0.46.

```
#One sample test for proportion
#Prest study, proportion is 25 out of total 65
#population proportion p =0.46
res <- prop.test(x = 25, n = 65, p = 0.46,
                 correct = FALSE)
# Printing the results
res
```

Output will be as follows:

```
        1-sample proportions test without continuity correction

data:  25 out of 65, null probability 0.46
X-squared = 1.5, df = 1, p-value = 0.2
alternative hypothesis: true p is not equal to 0.46
95 percent confidence interval:
 0.2760 0.5062
sample estimates:
       p
0.3846
```

Inference: The proportion of present study is not significantly different from the population proportion (p = 0.2).

29.3.10.2 Tests for Proportions: More than One Sample

Objective: To assess whether the proportions of alcoholics are significantly different between smoker and non-smoker

```
# #prop.test
mysmokers = mydata$ALCOHOLIC[mydata$SMOKER == '1']
mynonsmokers = mydata$ALCOHOLIC[mydata$SMOKER == '0']
x <- c(sum(mysmokers),sum(mynonsmokers))
n  <- c(length(mysmokers),length(mynonsmokers))

prop.test(x, n, p=NULL, alternative = "two.sided", conf.level = 0.99,
correct = FALSE)
```

Output will be as follows:

```
        2-sample test for equality of proportions without continuity
correction

data:  x out of n
X-squared = 19, df = 1, p-value = 1e-05
alternative hypothesis: two.sided
99 percent confidence interval:
 0.2651 0.9492
sample estimates:
 prop 1  prop 2
0.64286 0.03571
```

Inference: The proportions of alcoholics are significantly different ($p < 0.000$) between smoker and non-smoker.

29.3.11 Test for Inference: Categorical Data II (Also see Chap. 15)

29.3.11.1 Chi-Square Test

Objective: To find the association between smoking and alcohol intake

```
# #Chi-square test
mycrosstable <- table(mydata$SMOKER ,mydata$ALCOHOLIC )
dimnames(mycrosstable) <- list(smoker = c("N", "Y"), alcoholic =
c("N","Y"))
mycrosstable
(Xsq <- chisq.test(mycrosstable))  # Prints test summary
Xsq$observed    # observed counts (same as mycrosstable)
Xsq$expected    # expected counts under the null
Xsq$residuals   # Pearson residuals
Xsq$stdres      # standardized residuals
```

Output will be as follows:

```
> #Chi-square test

> mycrosstable <- table(mydata$SMOKER ,mydata$ALCOHOLIC )
> dimnames(mycrosstable) <- list(smoker = c("N", "Y"), alcoholic =
c("N","Y"))
> mycrosstable
         alcoholic
smoker   N   Y
     N  27   1
     Y   5   9
> (Xsq <- chisq.test(mycrosstable))  # Prints test summary

        Pearson's Chi-squared test with Yates' continuity correction

data:  mycrosstable
X-squared = 16, df = 1, p-value = 7e-05

> Xsq$observed    # observed counts (same as M)
         alcoholic
smoker   N   Y
     N  27   1
     Y   5   9
> Xsq$expected    # expected counts under the null
         alcoholic
smoker     N      Y
     N  21.33  6.667
     Y  10.67  3.333
> Xsq$residuals   # Pearson residuals
         alcoholic
smoker     N       Y
     N   1.227  -2.195
     Y  -1.735   3.104
> Xsq$stdres      # standardized residuals
         alcoholic
smoker     N       Y
     N   4.355  -4.355
     Y  -4.355   4.355
```

Inference: There is significant association between smoking and alcohol intake ($p < 0.001$).

29.3.11.2 Chi-Square Goodness of Fit

Objective: To find the goodness of fit of following data

Observed frequency	45	91	34	9
Expected frequency	100	100	100	100

```
##Chi-square goodess of fit
myobserved <- c(45, 91, 34, 9)
myexpprops <- c(1/4, 1/4, 1/4, 1/4)
res <- chisq.test(myobserved, p = myexpprops)
res
```

Output will be as follows:

```
> #Chi-square goodess of fit
> myobserved <- c(45, 91, 34, 9)
> myexpprops <- c(1/4, 1/4, 1/4, 1/4)
> res <- chisq.test(myobserved, p = myexpprops)
> res

        Chi-squared test for given probabilities

data:  myobserved
X-squared = 78.944, df = 3, p-value < 2.2e-16
```

Inference: Test for "Goodness of fit" of the data given is statistically significant ($p < 0.001$). Hence, observed population is significantly different from the expected population.

29.3.12 Test for Inference: Correlation and Regression (Also see Chap. 16)

29.3.12.1 Pearson's Correlation

Objective: To find the correlation between AGE and PASI score

```
# #Correlation test
cor.test(mydata$PASI, mydata$AGE, alternative = "two.sided",
         method = "pearson",
         exact = NULL, conf.level = 0.95, continuity = FALSE)
```

Output will be as follows:

```
        Pearson's product-moment correlation
data:  mydata$PASI and mydata$AGE
t = 1.9, df = 40, p-value = 0.06
alternative hypothesis: true correlation is not equal to 0
95 percent confidence interval:
 -0.01378  0.54687
sample estimates:
    cor
0.2914
```

Inference: The magnitude of correlation between AGE and PASI score is 0.2914.

29.3.12.2 Linear Regression

Objective: To assess the following model

$$PASI = (\beta_0 + \beta_1 AGE)$$

```
##Linear regression

mylm <- lm(PASI ~ AGE, data = mydata)

plot(PASI ~ AGE, data = mydata)
abline(coef(mylm))

summary(mylm)

anova(mylm)
```

Output will be as follows:

```
> #Linear regression
>
> summary(mylm)

Call:
lm(formula = PASI ~ AGE, data = mydata)

Residuals:
   Min     1Q Median     3Q    Max
-15.92  -9.96  -3.94   5.28  42.43

Coefficients:
            Estimate Std. Error t value Pr(>|t|)
(Intercept)    4.435      4.804    0.92    0.361
AGE            0.230      0.119    1.93    0.061 .
---
Signif. codes:  0 '***' 0.001 '**' 0.01 '*' 0.05 '.' 0.1 ' ' 1

Residual standard error: 13.8 on 40 degrees of freedom
Multiple R-squared:  0.0849,     Adjusted R-squared:  0.062
F-statistic: 3.71 on 1 and 40 DF,  p-value: 0.0612

> anova(mylm)
Analysis of Variance Table

Response: PASI
          Df Sum Sq Mean Sq F value Pr(>F)
AGE        1    706     706    3.71  0.061 .
Residuals 40   7606     190
---
Signif. codes:  0 '***' 0.001 '**' 0.01 '*' 0.05 '.' 0.1 ' ' 1
>
```

Inference: Model obtained

$$PASI = (4.435 + 0.23\, AGE)$$

Table 29.2 Results from a clinical laboratory test

Test result	Disease		Total
	Present	Absent	
Positive	231	32	263
Negative	27	54	81
Total	258	86	344

29.3.13 Analysis of Diagnostic Tests (Also see Chap. 20)

29.3.13.1 Sensitivity and Specificity (Requires Package "Caret")

Objective: Calculate the sensitivity and specificity of following data (Table 29.2).

```
##Sensitivity and Specificity
library(caret)
lvs <- c("normal", "abnormal")
truth <- factor(rep(lvs, times = c(86, 258)), levels = rev(lvs))
#Column totals
pred <- factor(
  c(
    rep(lvs, times = c(54, 32)),          # 'Normal Column'
    rep(lvs, times = c(27, 231))),        # 'Abnormal Column'
  levels = rev(lvs))

xtab <- table(pred, truth)

confusionMatrix(xtab)
```

Output will be as follows:

```
Confusion Matrix and Statistics

             truth
pred        abnormal normal
  abnormal       231     32
  normal          27     54

               Accuracy : 0.8285
                 95% CI : (0.7844, 0.8668)
    No Information Rate : 0.75
    P-Value [Acc > NIR] : 0.0003097

                  Kappa : 0.5336

 Mcnemar's Test P-Value : 0.6025370

            Sensitivity : 0.8953
            Specificity : 0.6279
         Pos Pred Value : 0.8783
         Neg Pred Value : 0.6667
             Prevalence : 0.7500
         Detection Rate : 0.6715
   Detection Prevalence : 0.7645
      Balanced Accuracy : 0.7616

       'Positive' Class : abnormal
```

Inference: Sensitivity $= 0.895$ and specificity $= 0.628$

29.3.14 Logistic Regression (Also see Chap. 28)

Objective: To assess the following model

$$p(\text{smokig habit, a binary variable}) = \frac{1}{1 + exp^{-(\beta_0 + \beta_1 AGE)}}$$

```
##Logistic regression
output1 <- glm(mydata$SMOKER ~ mydata$AGE, data=mydata,
family=binomial)
output1
```

Output will be as follows:

```
> #Logistic regression
> output1 <- glm(mydata$SMOKER ~ mydata$AGE, data=mydata,
family=binomial)
> output1

Call:  glm(formula = mydata$SMOKER ~ mydata$AGE, family = binomial,
    data = mydata)

Coefficients:
(Intercept)    mydata$AGE
   -1.1090        0.0113

Degrees of Freedom: 41 Total (i.e. Null);   40 Residual
Null Deviance:        53.5
Residual Deviance: 53.1    AIC: 57.1
>
```

Inference: Model obtained

$$p(\text{smokig habit, a binary variable}) = \frac{1}{1 + exp^{-(-1.11 + 0.113\ AGE)}}$$

29.3.15 Nonparametric tests (Also see Chap. 17)

29.3.15.1 Wilcoxon Test: One Sample
Objective: To find if median of AGE is significantly different from population
   median of 30

```
##Non parametric tests
median(mydata$AGE)

#One sample test- wilcoxon Signed Rank Test
wilcox.test(mydata$AGE, mu = 30, conf.int = TRUE)
```

Output will be as follows:

```
> median(mydata$AGE)
[1] 31.5
>
> wilcox.test(mydata$AGE, mu = 30, conf.int = TRUE)

    wilcoxon signed rank test with continuity correction

data:  mydata$AGE
V = 560, p-value = 0.1
alternative hypothesis: true location is not equal to 30
95 percent confidence interval:
 29.0 41.5
sample estimates:
(pseudo)median
             35
```

Inference: Median of AGE is not significantly different from median of population ($p = 0.1$).

29.3.15.2 Mann-Whitney-Wilcoxon Rank Sum Test

Objective: To find whether AGE is significantly different between SMOKER and NONSMOKER

```
#Mann-whitney-wilcoxon Rank Sum Test
wilcox.test(mydata$AGE[mydata$SMOKER == '0'],
mydata$AGE[mydata$SMOKER == '1'], mu = 0, conf.int = TRUE)

#Option II
wilcox.test(mydata$AGE ~ mydata$SMOKER)
```

Output will be as follows:

```
>
> wilcox.test(mydata$AGE[mydata$SMOKER == '0'],
mydata$AGE[mydata$SMOKER == '1'], mu = 0, conf.int = TRUE)

        Wilcoxon rank sum test with continuity correction

data:  mydata$AGE[mydata$SMOKER == "0"] and mydata$AGE[mydata$SMOKER
== "1"]
W = 156, p-value = 0.3
alternative hypothesis: true location shift is not equal to 0
95 percent confidence interval:
 -17    5
sample estimates:
difference in location
                     -5

        Wilcoxon rank sum test with continuity correction

data:  mydata$AGE by mydata$SMOKER
W = 156, p-value = 0.3
alternative hypothesis: true location shift is not equal to 0
```

Inference: AGE is not significantly different between SMOKER and NONSMOKER ($p = 0.3$).

29.3.15.3 Wilcoxon Signed Rank Test for Paired Observations

Objective: To assess the significance of difference between paired observations (y1, y2)

```
# wilcoxon signed rank test for paired observations
y1 <- c(21,34,67,54,89,37,12)
y2 <- c(19,30,65,49,79,35,9)

wilcox.test(y1, y2, paired=TRUE) # where y1 and y2 are numeric
```

Output will be as follows:

```
        Wilcoxon signed rank test with continuity correction

data:  y1 and y2
V = 28, p-value = 0.02
alternative hypothesis: true location shift is not equal to 0
```

Inference: Paired observations given are significantly different ($p = 0.02$).

29.3.15.4 Kruskal Wallis H-Test

Objective: To find significance of difference in PASI scores grouped according to
   STRESS

```
# Kruskal wallis test
kruskal.test(mydata$PASI ~ mydata$STRESS, data = mydata)
```

Output will be as follows:

```
> kruskal.test(mydata$PASI ~ mydata$STRESS, data = mydata)

       Kruskal-wallis rank sum test

data:   mydata$PASI by mydata$STRESS
Kruskal-wallis chi-squared = 11, df = 3, p-value = 0.01
```

Inference: PASI score is different among various STRESS scores ($p = 0.01$)

29.3.15.5 Spearman's Rank Correlation

Objective: To find correlation between AGE and PASI score

```
#Spearman's Rank Correlation test
cor.test(mydata$PASI, mydata$AGE, alternative = "two.sided",
         method = "spearman",
         exact = NULL, conf.level = 0.95, continuity = FALSE)
```

Output will be as follows:

```
        Spearman's rank correlation rho

data:   mydata$PASI and mydata$AGE
S = 10028, p-value = 0.2
alternative hypothesis: true rho is not equal to 0
sample estimates:
    rho
 0.1875
```

Inference: The magnitude of correlation between AGE and PASI score is $r = 0.1875$.

Appendices

Appendix 1: Area Under the Standard Normal Curve

z	0.00	0.01	0.02	0.03	0.04	0.05	0.06	0.07	0.08	0.09
0.0	0.00000	0.00399	0.00798	0.01197	0.01595	0.01994	0.02392	0.02790	0.03188	0.03586
0.1	0.03983	0.04380	0.04776	0.05172	0.05567	0.05962	0.06356	0.06749	0.07142	0.07535
0.2	0.07926	0.08317	0.08706	0.09095	0.09483	0.09871	0.10257	0.10642	0.11026	0.11409
0.3	0.11791	0.12172	0.12552	0.12930	0.13307	0.13683	0.14058	0.14431	0.14803	0.15173
0.4	0.15542	0.15910	0.16276	0.16640	0.17003	0.17364	0.17724	0.18082	0.18439	0.18793
0.5	0.19146	0.19497	0.19847	0.20194	0.20540	0.20884	0.21226	0.21566	0.21904	0.22240
0.6	0.22575	0.22907	0.23237	0.23565	0.23891	0.24215	0.24537	0.24857	0.25175	0.25490
0.7	0.25804	0.26115	0.26424	0.26730	0.27035	0.27337	0.27637	0.27935	0.28230	0.28524
0.8	0.28814	0.29103	0.29389	0.29673	0.29955	0.30234	0.30511	0.30785	0.31057	0.31327
0.9	0.31594	0.31859	0.32121	0.32381	0.32639	0.32894	0.33147	0.33398	0.33646	0.33891
1.0	0.34134	0.34375	0.34614	0.34849	0.35083	0.35314	0.35543	0.35769	0.35993	0.36214
1.1	0.36433	0.36650	0.36864	0.37076	0.37286	0.37493	0.37698	0.37900	0.38100	0.38298
1.2	0.38493	0.38686	0.38877	0.39065	0.39251	0.39435	0.39617	0.39796	0.39973	0.40147
1.3	0.40320	0.40490	0.40658	0.40824	0.40988	0.41149	0.41308	0.41466	0.41621	0.41774
1.4	0.41924	0.42073	0.42220	0.42364	0.42507	0.42647	0.42785	0.42922	0.43056	0.43189
1.5	0.43319	0.43448	0.43574	0.43699	0.43822	0.43943	0.44062	0.44179	0.44295	0.44408
1.6	0.44520	0.44630	0.44738	0.44845	0.44950	0.45053	0.45154	0.45254	0.45352	0.45449
1.7	0.45543	0.45637	0.45728	0.45818	0.45907	0.45994	0.46080	0.46164	0.46246	0.46327
1.8	0.46407	0.46485	0.46562	0.46638	0.46712	0.46784	0.46856	0.46926	0.46995	0.47062
1.9	0.47128	0.47193	0.47257	0.47320	0.47381	0.47441	0.47500	0.47558	0.47615	0.47670
2.0	0.47725	0.47778	0.47831	0.47882	0.47932	0.47982	0.48030	0.48077	0.48124	0.48169
2.1	0.48214	0.48257	0.48300	0.48341	0.48382	0.48422	0.48461	0.48500	0.48537	0.48574
2.2	0.48610	0.48645	0.48679	0.48713	0.48745	0.48778	0.48809	0.48840	0.48870	0.48899
2.3	0.48928	0.48956	0.48983	0.49010	0.49036	0.49061	0.49086	0.49111	0.49134	0.49158

	0.00	0.01	0.02	0.03	0.04	0.05	0.06	0.07	0.08	0.09
2.4	0.49180	0.49202	0.49224	0.49245	0.49266	0.49286	0.49305	0.49324	0.49343	0.49361
2.5	0.49379	0.49396	0.49413	0.49430	0.49446	0.49461	0.49477	0.49492	0.49506	0.49520
2.6	0.49534	0.49547	0.49560	0.49573	0.49585	0.49598	0.49609	0.49621	0.49632	0.49643
2.7	0.49653	0.49664	0.49674	0.49683	0.49693	0.49702	0.49711	0.49720	0.49728	0.49736
2.8	0.49744	0.49752	0.49760	0.49767	0.49774	0.49781	0.49788	0.49795	0.49801	0.49807
2.9	0.49813	0.49819	0.49825	0.49831	0.49836	0.49841	0.49846	0.49851	0.49856	0.49861
3.0	0.49865	0.49869	0.49874	0.49878	0.49882	0.49886	0.49889	0.49893	0.49896	0.49900
3.1	0.49903	0.49906	0.49910	0.49913	0.49916	0.49918	0.49921	0.49924	0.49926	0.49929
3.2	0.49931	0.49934	0.49936	0.49938	0.49940	0.49942	0.49944	0.49946	0.49948	0.49950
3.3	0.49952	0.49953	0.49955	0.49957	0.49958	0.49960	0.49961	0.49962	0.49964	0.49965
3.4	0.49966	0.49968	0.49969	0.49970	0.49971	0.49972	0.49973	0.49974	0.49975	0.49976
3.5	0.49977	0.49978	0.49978	0.49979	0.49980	0.49981	0.49981	0.49982	0.49983	0.49983
3.6	0.49984	0.49985	0.49985	0.49986	0.49986	0.49987	0.49987	0.49988	0.49988	0.49989
3.7	0.49989	0.49990	0.49990	0.49990	0.49991	0.49991	0.49992	0.49992	0.49992	0.49992
3.8	0.49993	0.49993	0.49993	0.49994	0.49994	0.49994	0.49994	0.49995	0.49995	0.49995
3.9	0.49995	0.49995	0.49996	0.49996	0.49996	0.49996	0.49996	0.49996	0.49997	0.49997
4.0	0.49997	0.49997	0.49997	0.49997	0.49997	0.49997	0.49998	0.49998	0.49998	0.49998

Printed with permission from NIST/SEMATECH e-Handbook of Statistical Methods, http://www.itl.nist.gov/div898/handbook/, 12 May 2019

Note: Value given are area under the curve before z value in the right half of the standard normal distribution. The area after the z value $[p \ (X>z)]$ can be estimated by subtracting the probabilities given above from 0.5

Appendix 2: Distribution of "t"—Probability Levels

v	0.90	0.95	0.975	0.99	0.995	0.999
1	3.078	6.314	12.706	31.821	63.657	318.313
2	1.886	2.920	4.303	6.965	9.925	22.327
3	1.638	2.353	3.182	4.541	5.841	10.215
4	1.533	2.132	2.776	3.747	4.604	7.173
5	1.476	2.015	2.571	3.365	4.032	5.893
6	1.440	1.943	2.447	3.143	3.707	5.208
7	1.415	1.895	2.365	2.998	3.499	4.782
8	1.397	1.860	2.306	2.896	3.355	4.499
9	1.383	1.833	2.262	2.821	3.250	4.296
10	1.372	1.812	2.228	2.764	3.169	4.143
11	1.363	1.796	2.201	2.718	3.106	4.024
12	1.356	1.782	2.179	2.681	3.055	3.929
13	1.350	1.771	2.160	2.650	3.012	3.852
14	1.345	1.761	2.145	2.624	2.977	3.787
15	1.341	1.753	2.131	2.602	2.947	3.733
16	1.337	1.746	2.120	2.583	2.921	3.686
17	1.333	1.740	2.110	2.567	2.898	3.646
18	1.330	1.734	2.101	2.552	2.878	3.610
19	1.328	1.729	2.093	2.539	2.861	3.579
20	1.325	1.725	2.086	2.528	2.845	3.552
21	1.323	1.721	2.080	2.518	2.831	3.527
22	1.321	1.717	2.074	2.508	2.819	3.505

23	1.319	1.714	2.069	2.500	2.807	3.485
24	1.318	1.711	2.064	2.492	2.797	3.467
25	1.316	1.708	2.060	2.485	2.787	3.450
26	1.315	1.706	2.056	2.479	2.779	3.435
27	1.314	1.703	2.052	2.473	2.771	3.421
28	1.313	1.701	2.048	2.467	2.763	3.408
29	1.311	1.699	2.045	2.462	2.756	3.396
30	1.310	1.697	2.042	2.457	2.750	3.385
31	1.309	1.696	2.040	2.453	2.744	3.375
32	1.309	1.694	2.037	2.449	2.738	3.365
33	1.308	1.692	2.035	2.445	2.733	3.356
34	1.307	1.691	2.032	2.441	2.728	3.348
35	1.306	1.690	2.030	2.438	2.724	3.340
36	1.306	1.688	2.028	2.434	2.719	3.333
37	1.305	1.687	2.026	2.431	2.715	3.326
38	1.304	1.686	2.024	2.429	2.712	3.319
39	1.304	1.685	2.023	2.426	2.708	3.313
40	1.303	1.684	2.021	2.423	2.704	3.307
41	1.303	1.683	2.020	2.421	2.701	3.301
42	1.302	1.682	2.018	2.418	2.698	3.296
43	1.302	1.681	2.017	2.416	2.695	3.291
44	1.301	1.680	2.015	2.414	2.692	3.286
45	1.301	1.679	2.014	2.412	2.690	3.281
46	1.300	1.679	2.013	2.410	2.687	3.277
47	1.300	1.678	2.012	2.408	2.685	3.273
48	1.299	1.677	2.011	2.407	2.682	3.269

(continued)

49	1.299	1.677	2.010	2.405	2.680	3.265
50	1.299	1.676	2.009	2.403	2.678	3.261
60	1.296	1.671	2.000	2.390	2.660	3.232
70	1.294	1.667	1.994	2.381	2.648	3.211
80	1.292	1.664	1.990	2.374	2.639	3.195
90	1.291	1.662	1.987	2.368	2.632	3.183
100	1.290	1.660	1.984	2.364	2.626	3.174
∞	1.282	1.645	1.960	2.326	2.576	3.090

Printed with permission from NIST/SEMATECH e-Handbook of Statistical Methods, http://www.itl.nist.gov/div898/handbook/, 12 May 2019

Note: To assess the significance of 5% level for one-tailed test, use 0.95; for two-tailed test, use 0.975

Appendix 3: Distribution of Chi-Square (χ^2) – Probability Levels

Upper-tail critical values of chi-square distribution with ν degrees of freedom
Probability less than the critical value

ν	0.90	0.95	0.975	0.99	0.999
1	2.706	3.841	5.024	6.635	10.828
2	4.605	5.991	7.378	9.210	13.816
3	6.251	7.815	9.348	11.345	16.266
4	7.779	9.488	11.143	13.277	18.467
5	9.236	11.070	12.833	15.086	20.515
6	10.645	12.592	14.449	16.812	22.458
7	12.017	14.067	16.013	18.475	24.322
8	13.362	15.507	17.535	20.090	26.125
9	14.684	16.919	19.023	21.666	27.877
10	15.987	18.307	20.483	23.209	29.588
11	17.275	19.675	21.920	24.725	31.264
12	18.549	21.026	23.337	26.217	32.910
13	19.812	22.362	24.736	27.688	34.528
14	21.064	23.685	26.119	29.141	36.123
15	22.307	24.996	27.488	30.578	37.697
16	23.542	26.296	28.845	32.000	39.252
17	24.769	27.587	30.191	33.409	40.790
18	25.989	28.869	31.526	34.805	42.312
19	27.204	30.144	32.852	36.191	43.820
20	28.412	31.410	34.170	37.566	45.315

(continued)

21	29.615	32.671	35.479	38.932	46.797
22	30.813	33.924	36.781	40.289	48.268
23	32.007	35.172	38.076	41.638	49.728
24	33.196	36.415	39.364	42.980	51.179
25	34.382	37.652	40.646	44.314	52.620
26	35.563	38.885	41.923	45.642	54.052
27	36.741	40.113	43.195	46.963	55.476
28	37.916	41.337	44.461	48.278	56.892
29	39.087	42.557	45.722	49.588	58.301
30	40.256	43.773	46.979	50.892	59.703
31	41.422	44.985	48.232	52.191	61.098
32	42.585	46.194	49.480	53.486	62.487
33	43.745	47.400	50.725	54.776	63.870
34	44.903	48.602	51.966	56.061	65.247
35	46.059	49.802	53.203	57.342	66.619
36	47.212	50.998	54.437	58.619	67.985
37	48.363	52.192	55.668	59.893	69.347
38	49.513	53.384	56.896	61.162	70.703
39	50.660	54.572	58.120	62.428	72.055
40	51.805	55.758	59.342	63.691	73.402
41	52.949	56.942	60.561	64.950	74.745
42	54.090	58.124	61.777	66.206	76.084
43	55.230	59.304	62.990	67.459	77.419
44	56.369	60.481	64.201	68.710	78.750
45	57.505	61.656	65.410	69.957	80.077
46	58.641	62.830	66.617	71.201	81.400
47	59.774	64.001	67.821	72.443	82.720
48	60.907	65.171	69.023	73.683	84.037

76	92.166	97.351	101.999	107.583	119.850
77	93.270	98.484	103.158	108.771	121.100
78	94.374	99.617	104.316	109.958	122.348
79	95.476	100.749	105.473	111.144	123.594
80	96.578	101.879	106.629	112.329	124.839
81	97.680	103.010	107.783	113.512	126.083
82	98.780	104.139	108.937	114.695	127.324
83	99.880	105.267	110.090	115.876	128.565
84	100.980	106.395	111.242	117.057	129.804
85	102.079	107.522	112.393	118.236	131.041
86	103.177	108.648	113.544	119.414	132.277
87	104.275	109.773	114.693	120.591	133.512
88	105.372	110.898	115.841	121.767	134.746
89	106.469	112.022	116.989	122.942	135.978
90	107.565	113.145	118.136	124.116	137.208
91	108.661	114.268	119.282	125.289	138.438
92	109.756	115.390	120.427	126.462	139.666
93	110.850	116.511	121.571	127.633	140.893
94	111.944	117.632	122.715	128.803	142.119
95	113.038	118.752	123.858	129.973	143.344
96	114.131	119.871	125.000	131.141	144.567
97	115.223	120.990	126.141	132.309	145.789
98	116.315	122.108	127.282	133.476	147.010
99	117.407	123.225	128.422	134.642	148.230
100	118.498	124.342	129.561	135.807	149.449
100	118.498	124.342	129.561	135.807	149.449

Appendix 4: F-Test Table – 5% Point (1–10)

	1	2	3	4	5	6	7	8	9	10
1	161.448	199.500	215.707	224.583	230.162	233.986	236.768	238.882	240.543	241.882
2	18.513	19.000	19.164	19.247	19.296	19.330	19.353	19.371	19.385	19.396
3	10.128	9.552	9.277	9.117	9.013	8.941	8.887	8.845	8.812	8.786
4	7.709	6.944	6.591	6.388	6.256	6.163	6.094	6.041	5.999	5.964
5	6.608	5.786	5.409	5.192	5.050	4.950	4.876	4.818	4.772	4.735
6	5.987	5.143	4.757	4.534	4.387	4.284	4.207	4.147	4.099	4.060
7	5.591	4.737	4.347	4.120	3.972	3.866	3.787	3.726	3.677	3.637
8	5.318	4.459	4.066	3.838	3.687	3.581	3.500	3.438	3.388	3.347
9	5.117	4.256	3.863	3.633	3.482	3.374	3.293	3.230	3.179	3.137
10	4.965	4.103	3.708	3.478	3.326	3.217	3.135	3.072	3.020	2.978
11	4.844	3.982	3.587	3.357	3.204	3.095	3.012	2.948	2.896	2.854
12	4.747	3.885	3.490	3.259	3.106	2.996	2.913	2.849	2.796	2.753
13	4.667	3.806	3.411	3.179	3.025	2.915	2.832	2.767	2.714	2.671
14	4.600	3.739	3.344	3.112	2.958	2.848	2.764	2.699	2.646	2.602
15	4.543	3.682	3.287	3.056	2.901	2.790	2.707	2.641	2.588	2.544
16	4.494	3.634	3.239	3.007	2.852	2.741	2.657	2.591	2.538	2.494
17	4.451	3.592	3.197	2.965	2.810	2.699	2.614	2.548	2.494	2.450
18	4.414	3.555	3.160	2.928	2.773	2.661	2.577	2.510	2.456	2.412
19	4.381	3.522	3.127	2.895	2.740	2.628	2.544	2.477	2.423	2.378
20	4.351	3.493	3.098	2.866	2.711	2.599	2.514	2.447	2.393	2.348
21	4.325	3.467	3.072	2.840	2.685	2.573	2.488	2.420	2.366	2.321
22	4.301	3.443	3.049	2.817	2.661	2.549	2.464	2.397	2.342	2.297
23	4.279	3.422	3.028	2.796	2.640	2.528	2.442	2.375	2.320	2.275

(continued)

24	4.260	3.403	3.009	2.776	2.621	2.508	2.423	2.355	2.300	2.255
25	4.242	3.385	2.991	2.759	2.603	2.490	2.405	2.337	2.282	2.236
26	4.225	3.369	2.975	2.743	2.587	2.474	2.388	2.321	2.265	2.220
27	4.210	3.354	2.960	2.728	2.572	2.459	2.373	2.305	2.250	2.204
28	4.196	3.340	2.947	2.714	2.558	2.445	2.359	2.291	2.236	2.190
29	4.183	3.328	2.934	2.701	2.545	2.432	2.346	2.278	2.223	2.177
30	4.171	3.316	2.922	2.690	2.534	2.421	2.334	2.266	2.211	2.165
31	4.160	3.305	2.911	2.679	2.523	2.409	2.323	2.255	2.199	2.153
32	4.149	3.295	2.901	2.668	2.512	2.399	2.313	2.244	2.189	2.142
33	4.139	3.285	2.892	2.659	2.503	2.389	2.303	2.235	2.179	2.133
34	4.130	3.276	2.883	2.650	2.494	2.380	2.294	2.225	2.170	2.123
35	4.121	3.267	2.874	2.641	2.485	2.372	2.285	2.217	2.161	2.114
36	4.113	3.259	2.866	2.634	2.477	2.364	2.277	2.209	2.153	2.106
37	4.105	3.252	2.859	2.626	2.470	2.356	2.270	2.201	2.145	2.098
38	4.098	3.245	2.852	2.619	2.463	2.349	2.262	2.194	2.138	2.091
39	4.091	3.238	2.845	2.612	2.456	2.342	2.255	2.187	2.131	2.084
40	4.085	3.232	2.839	2.606	2.449	2.336	2.249	2.180	2.124	2.077
41	4.079	3.226	2.833	2.600	2.443	2.330	2.243	2.174	2.118	2.071
42	4.073	3.220	2.827	2.594	2.438	2.324	2.237	2.168	2.112	2.065
43	4.067	3.214	2.822	2.589	2.432	2.318	2.232	2.163	2.106	2.059
44	4.062	3.209	2.816	2.584	2.427	2.313	2.226	2.157	2.101	2.054
45	4.057	3.204	2.812	2.579	2.422	2.308	2.221	2.152	2.096	2.049
46	4.052	3.200	2.807	2.574	2.417	2.304	2.216	2.147	2.091	2.044
47	4.047	3.195	2.802	2.570	2.413	2.299	2.212	2.143	2.086	2.039
48	4.043	3.191	2.798	2.565	2.409	2.295	2.207	2.138	2.082	2.035
49	4.038	3.187	2.794	2.561	2.404	2.290	2.203	2.134	2.077	2.030
50	4.034	3.183	2.790	2.557	2.400	2.286	2.199	2.130	2.073	2.026
51	4.030	3.179	2.786	2.553	2.397	2.283	2.195	2.126	2.069	2.022

52	4.027	3.175	2.783	2.550	2.393	2.279	2.192	2.122	2.066	2.018
53	4.023	3.172	2.779	2.546	2.389	2.275	2.188	2.119	2.062	2.015
54	4.020	3.168	2.776	2.543	2.386	2.272	2.185	2.115	2.059	2.011
55	4.016	3.165	2.773	2.540	2.383	2.269	2.181	2.112	2.055	2.008
56	4.013	3.162	2.769	2.537	2.380	2.266	2.178	2.109	2.052	2.005
57	4.010	3.159	2.766	2.534	2.377	2.263	2.175	2.106	2.049	2.001
58	4.007	3.156	2.764	2.531	2.374	2.260	2.172	2.103	2.046	1.998
59	4.004	3.153	2.761	2.528	2.371	2.257	2.169	2.100	2.043	1.995
60	4.001	3.150	2.758	2.525	2.368	2.254	2.167	2.097	2.040	1.993
61	3.998	3.148	2.755	2.523	2.366	2.251	2.164	2.094	2.037	1.990
62	3.996	3.145	2.753	2.520	2.363	2.249	2.161	2.092	2.035	1.987
63	3.993	3.143	2.751	2.518	2.361	2.246	2.159	2.089	2.032	1.985
64	3.991	3.140	2.748	2.515	2.358	2.244	2.156	2.087	2.030	1.982
65	3.989	3.138	2.746	2.513	2.356	2.242	2.154	2.084	2.027	1.980
66	3.986	3.136	2.744	2.511	2.354	2.239	2.152	2.082	2.025	1.977
67	3.984	3.134	2.742	2.509	2.352	2.237	2.150	2.080	2.023	1.975
68	3.982	3.132	2.740	2.507	2.350	2.235	2.148	2.078	2.021	1.973
69	3.980	3.130	2.737	2.505	2.348	2.233	2.145	2.076	2.019	1.971
70	3.978	3.128	2.736	2.503	2.346	2.231	2.143	2.074	2.017	1.969
71	3.976	3.126	2.734	2.501	2.344	2.229	2.142	2.072	2.015	1.967
72	3.974	3.124	2.732	2.499	2.342	2.227	2.140	2.070	2.013	1.965
73	3.972	3.122	2.730	2.497	2.340	2.226	2.138	2.068	2.011	1.963
74	3.970	3.120	2.728	2.495	2.338	2.224	2.136	2.066	2.009	1.961
75	3.968	3.119	2.727	2.494	2.337	2.222	2.134	2.064	2.007	1.959
76	3.967	3.117	2.725	2.492	2.335	2.220	2.133	2.063	2.006	1.958
77	3.965	3.115	2.723	2.490	2.333	2.219	2.131	2.061	2.004	1.956
78	3.963	3.114	2.722	2.489	2.332	2.217	2.129	2.059	2.002	1.954

(continued)

79	3.962	3.112	2.720	2.487	2.330	2.216	2.128	2.058	2.001	1.953
80	3.960	3.111	2.719	2.486	2.329	2.214	2.126	2.056	1.999	1.951
81	3.959	3.109	2.717	2.484	2.327	2.213	2.125	2.055	1.998	1.950
82	3.957	3.108	2.716	2.483	2.326	2.211	2.123	2.053	1.996	1.948
83	3.956	3.107	2.715	2.482	2.324	2.210	2.122	2.052	1.995	1.947
84	3.955	3.105	2.713	2.480	2.323	2.209	2.121	2.051	1.993	1.945
85	3.953	3.104	2.712	2.479	2.322	2.207	2.119	2.049	1.992	1.944
86	3.952	3.103	2.711	2.478	2.321	2.206	2.118	2.048	1.991	1.943
87	3.951	3.101	2.709	2.476	2.319	2.205	2.117	2.047	1.989	1.941
88	3.949	3.100	2.708	2.475	2.318	2.203	2.115	2.045	1.988	1.940
89	3.948	3.099	2.707	2.474	2.317	2.202	2.114	2.044	1.987	1.939
90	3.947	3.098	2.706	2.473	2.316	2.201	2.113	2.043	1.986	1.938
91	3.946	3.097	2.705	2.472	2.315	2.200	2.112	2.042	1.984	1.936
92	3.945	3.095	2.704	2.471	2.313	2.199	2.111	2.041	1.983	1.935
93	3.943	3.094	2.703	2.470	2.312	2.198	2.110	2.040	1.982	1.934
94	3.942	3.093	2.701	2.469	2.311	2.197	2.109	2.038	1.981	1.933
95	3.941	3.092	2.700	2.467	2.310	2.196	2.108	2.037	1.980	1.932
96	3.940	3.091	2.699	2.466	2.309	2.195	2.106	2.036	1.979	1.931
97	3.939	3.090	2.698	2.465	2.308	2.194	2.105	2.035	1.978	1.930
98	3.938	3.089	2.697	2.465	2.307	2.193	2.104	2.034	1.977	1.929
99	3.937	3.088	2.696	2.464	2.306	2.192	2.103	2.033	1.976	1.928
100	3.936	3.087	2.696	2.463	2.305	2.191	2.103	2.032	1.975	1.927

Printed with permission from NIST/SEMATECH e-Handbook of Statistical Methods, http://www.itl.nist.gov/div898/handbook/, 12 May 2019
Note: *df* for higher variance should be looked row wise

F-Test Table – 5% Point (11–20)

	11	12	13	14	15	16	17	18	19	20
1	242.983	243.906	244.690	245.364	245.950	246.464	246.918	247.323	247.686	248.013
2	19.405	19.413	19.419	19.424	19.429	19.433	19.437	19.440	19.443	19.446
3	8.763	8.745	8.729	8.715	8.703	8.692	8.683	8.675	8.667	8.660
4	5.936	5.912	5.891	5.873	5.858	5.844	5.832	5.821	5.811	5.803
5	4.704	4.678	4.655	4.636	4.619	4.604	4.590	4.579	4.568	4.558
6	4.027	4.000	3.976	3.956	3.938	3.922	3.908	3.896	3.884	3.874
7	3.603	3.575	3.550	3.529	3.511	3.494	3.480	3.467	3.455	3.445
8	3.313	3.284	3.259	3.237	3.218	3.202	3.187	3.173	3.161	3.150
9	3.102	3.073	3.048	3.025	3.006	2.989	2.974	2.960	2.948	2.936
10	2.943	2.913	2.887	2.865	2.845	2.828	2.812	2.798	2.785	2.774
11	2.818	2.788	2.761	2.739	2.719	2.701	2.685	2.671	2.658	2.646
12	2.717	2.687	2.660	2.637	2.617	2.599	2.583	2.568	2.555	2.544
13	2.635	2.604	2.577	2.554	2.533	2.515	2.499	2.484	2.471	2.459
14	2.565	2.534	2.507	2.484	2.463	2.445	2.428	2.413	2.400	2.388
15	2.507	2.475	2.448	2.424	2.403	2.385	2.368	2.353	2.340	2.328
16	2.456	2.425	2.397	2.373	2.352	2.333	2.317	2.302	2.288	2.276
17	2.413	2.381	2.353	2.329	2.308	2.289	2.272	2.257	2.243	2.230
18	2.374	2.342	2.314	2.290	2.269	2.250	2.233	2.217	2.203	2.191
19	2.340	2.308	2.280	2.256	2.234	2.215	2.198	2.182	2.168	2.155
20	2.310	2.278	2.250	2.225	2.203	2.184	2.167	2.151	2.137	2.124
21	2.283	2.250	2.222	2.197	2.176	2.156	2.139	2.123	2.109	2.096
22	2.259	2.226	2.198	2.173	2.151	2.131	2.114	2.098	2.084	2.071
23	2.236	2.204	2.175	2.150	2.128	2.109	2.091	2.075	2.061	2.048

(continued)

24	2.216	2.183	2.155	2.130	2.108	2.088	2.070	2.054	2.040	2.027
25	2.198	2.165	2.136	2.111	2.089	2.069	2.051	2.035	2.021	2.007
26	2.181	2.148	2.119	2.094	2.072	2.052	2.034	2.018	2.003	1.990
27	2.166	2.132	2.103	2.078	2.056	2.036	2.018	2.002	1.987	1.974
28	2.151	2.118	2.089	2.064	2.041	2.021	2.003	1.987	1.972	1.959
29	2.138	2.104	2.075	2.050	2.027	2.007	1.989	1.973	1.958	1.945
30	2.126	2.092	2.063	2.037	2.015	1.995	1.976	1.960	1.945	1.932
31	2.114	2.080	2.051	2.026	2.003	1.983	1.965	1.948	1.933	1.920
32	2.103	2.070	2.040	2.015	1.992	1.972	1.953	1.937	1.922	1.908
33	2.093	2.060	2.030	2.004	1.982	1.961	1.943	1.926	1.911	1.898
34	2.084	2.050	2.021	1.995	1.972	1.952	1.933	1.917	1.902	1.888
35	2.075	2.041	2.012	1.986	1.963	1.942	1.924	1.907	1.892	1.878
36	2.067	2.033	2.003	1.977	1.954	1.934	1.915	1.899	1.883	1.870
37	2.059	2.025	1.995	1.969	1.946	1.926	1.907	1.890	1.875	1.861
38	2.051	2.017	1.988	1.962	1.939	1.918	1.899	1.883	1.867	1.853
39	2.044	2.010	1.981	1.954	1.931	1.911	1.892	1.875	1.860	1.846
40	2.038	2.003	1.974	1.948	1.924	1.904	1.885	1.868	1.853	1.839
41	2.031	1.997	1.967	1.941	1.918	1.897	1.879	1.862	1.846	1.832
42	2.025	1.991	1.961	1.935	1.912	1.891	1.872	1.855	1.840	1.826
43	2.020	1.985	1.955	1.929	1.906	1.885	1.866	1.849	1.834	1.820
44	2.014	1.980	1.950	1.924	1.900	1.879	1.861	1.844	1.828	1.814
45	2.009	1.974	1.945	1.918	1.895	1.874	1.855	1.838	1.823	1.808
46	2.004	1.969	1.940	1.913	1.890	1.869	1.850	1.833	1.817	1.803
47	1.999	1.965	1.935	1.908	1.885	1.864	1.845	1.828	1.812	1.798
48	1.995	1.960	1.930	1.904	1.880	1.859	1.840	1.823	1.807	1.793
49	1.990	1.956	1.926	1.899	1.876	1.855	1.836	1.819	1.803	1.789
50	1.986	1.952	1.921	1.895	1.871	1.850	1.831	1.814	1.798	1.784
51	1.982	1.947	1.917	1.891	1.867	1.846	1.827	1.810	1.794	1.780

52	1.978	1.944	1.913	1.887	1.863	1.842	1.823	1.806	1.790	1.776
53	1.975	1.940	1.910	1.883	1.859	1.838	1.819	1.802	1.786	1.772
54	1.971	1.936	1.906	1.879	1.856	1.835	1.816	1.798	1.782	1.768
55	1.968	1.933	1.903	1.876	1.852	1.831	1.812	1.795	1.779	1.764
56	1.964	1.930	1.899	1.873	1.849	1.828	1.809	1.791	1.775	1.761
57	1.961	1.926	1.896	1.869	1.846	1.824	1.805	1.788	1.772	1.757
58	1.958	1.923	1.893	1.866	1.842	1.821	1.802	1.785	1.769	1.754
59	1.955	1.920	1.890	1.863	1.839	1.818	1.799	1.781	1.766	1.751
60	1.952	1.917	1.887	1.860	1.836	1.815	1.796	1.778	1.763	1.748
61	1.949	1.915	1.884	1.857	1.834	1.812	1.793	1.776	1.760	1.745
62	1.947	1.912	1.882	1.855	1.831	1.809	1.790	1.773	1.757	1.742
63	1.944	1.909	1.879	1.852	1.828	1.807	1.787	1.770	1.754	1.739
64	1.942	1.907	1.876	1.849	1.826	1.804	1.785	1.767	1.751	1.737
65	1.939	1.904	1.874	1.847	1.823	1.802	1.782	1.765	1.749	1.734
66	1.937	1.902	1.871	1.845	1.821	1.799	1.780	1.762	1.746	1.732
67	1.935	1.900	1.869	1.842	1.818	1.797	1.777	1.760	1.744	1.729
68	1.932	1.897	1.867	1.840	1.816	1.795	1.775	1.758	1.742	1.727
69	1.930	1.895	1.865	1.838	1.814	1.792	1.773	1.755	1.739	1.725
70	1.928	1.893	1.863	1.836	1.812	1.790	1.771	1.753	1.737	1.722
71	1.926	1.891	1.861	1.834	1.810	1.788	1.769	1.751	1.735	1.720
72	1.924	1.889	1.859	1.832	1.808	1.786	1.767	1.749	1.733	1.718
73	1.922	1.887	1.857	1.830	1.806	1.784	1.765	1.747	1.731	1.716
74	1.921	1.885	1.855	1.828	1.804	1.782	1.763	1.745	1.729	1.714
75	1.919	1.884	1.853	1.826	1.802	1.780	1.761	1.743	1.727	1.712
76	1.917	1.882	1.851	1.824	1.800	1.778	1.759	1.741	1.725	1.710
77	1.915	1.880	1.849	1.822	1.798	1.777	1.757	1.739	1.723	1.708
78	1.914	1.878	1.848	1.821	1.797	1.775	1.755	1.738	1.721	1.707

(continued)

79	1.912	1.877	1.846	1.819	1.795	1.773	1.754	1.736	1.720	1.705
80	1.910	1.875	1.845	1.817	1.793	1.772	1.752	1.734	1.718	1.703
81	1.909	1.874	1.843	1.816	1.792	1.770	1.750	1.733	1.716	1.702
82	1.907	1.872	1.841	1.814	1.790	1.768	1.749	1.731	1.715	1.700
83	1.906	1.871	1.840	1.813	1.789	1.767	1.747	1.729	1.713	1.698
84	1.905	1.869	1.838	1.811	1.787	1.765	1.746	1.728	1.712	1.697
85	1.903	1.868	1.837	1.810	1.786	1.764	1.744	1.726	1.710	1.695
86	1.902	1.867	1.836	1.808	1.784	1.762	1.743	1.725	1.709	1.694
87	1.900	1.865	1.834	1.807	1.783	1.761	1.741	1.724	1.707	1.692
88	1.899	1.864	1.833	1.806	1.782	1.760	1.740	1.722	1.706	1.691
89	1.898	1.863	1.832	1.804	1.780	1.758	1.739	1.721	1.705	1.690
90	1.897	1.861	1.830	1.803	1.779	1.757	1.737	1.720	1.703	1.688
91	1.895	1.860	1.829	1.802	1.778	1.756	1.736	1.718	1.702	1.687
92	1.894	1.859	1.828	1.801	1.776	1.755	1.735	1.717	1.701	1.686
93	1.893	1.858	1.827	1.800	1.775	1.753	1.734	1.716	1.699	1.684
94	1.892	1.857	1.826	1.798	1.774	1.752	1.733	1.715	1.698	1.683
95	1.891	1.856	1.825	1.797	1.773	1.751	1.731	1.713	1.697	1.682
96	1.890	1.854	1.823	1.796	1.772	1.750	1.730	1.712	1.696	1.681
97	1.889	1.853	1.822	1.795	1.771	1.749	1.729	1.711	1.695	1.680
98	1.888	1.852	1.821	1.794	1.770	1.748	1.728	1.710	1.694	1.679
99	1.887	1.851	1.820	1.793	1.769	1.747	1.727	1.709	1.693	1.678
100	1.886	1.850	1.819	1.792	1.768	1.746	1.726	1.708	1.691	1.676

Printed with permission from NIST/SEMATECH e-Handbook of Statistical Methods, http://www.itl.nist.gov/div898/handbook/, 12 May 2019

Note: df for higher variance should be looked row wise

Appendix 5: F-Test Table – 1% Point (1–10)

	1	2	3	4	5	6	7	8	9	10
1	4052.19	4999.52	5403.34	5624.62	5763.65	5858.97	5928.33	5981.10	6022.50	6055.85
2	98.502	99.000	99.166	99.249	99.300	99.333	99.356	99.374	99.388	99.399
3	34.116	30.816	29.457	28.710	28.237	27.911	27.672	27.489	27.345	27.229
4	21.198	18.000	16.694	15.977	15.522	15.207	14.976	14.799	14.659	14.546
5	16.258	13.274	12.060	11.392	10.967	10.672	10.456	10.289	10.158	10.051
6	13.745	10.925	9.780	9.148	8.746	8.466	8.260	8.102	7.976	7.874
7	12.246	9.547	8.451	7.847	7.460	7.191	6.993	6.840	6.719	6.620
8	11.259	8.649	7.591	7.006	6.632	6.371	6.178	6.029	5.911	5.814
9	10.561	8.022	6.992	6.422	6.057	5.802	5.613	5.467	5.351	5.257
10	10.044	7.559	6.552	5.994	5.636	5.386	5.200	5.057	4.942	4.849
11	9.646	7.206	6.217	5.668	5.316	5.069	4.886	4.744	4.632	4.539
12	9.330	6.927	5.953	5.412	5.064	4.821	4.640	4.499	4.388	4.296
13	9.074	6.701	5.739	5.205	4.862	4.620	4.441	4.302	4.191	4.100
14	8.862	6.515	5.564	5.035	4.695	4.456	4.278	4.140	4.030	3.939
15	8.683	6.359	5.417	4.893	4.556	4.318	4.142	4.004	3.895	3.805
16	8.531	6.226	5.292	4.773	4.437	4.202	4.026	3.890	3.780	3.691
17	8.400	6.112	5.185	4.669	4.336	4.102	3.927	3.791	3.682	3.593
18	8.285	6.013	5.092	4.579	4.248	4.015	3.841	3.705	3.597	3.508
19	8.185	5.926	5.010	4.500	4.171	3.939	3.765	3.631	3.523	3.434
20	8.096	5.849	4.938	4.431	4.103	3.871	3.699	3.564	3.457	3.368
21	8.017	5.780	4.874	4.369	4.042	3.812	3.640	3.506	3.398	3.310
22	7.945	5.719	4.817	4.313	3.988	3.758	3.587	3.453	3.346	3.258
23	7.881	5.664	4.765	4.264	3.939	3.710	3.539	3.406	3.299	3.211

(continued)

24	3.168	3.256	3.363	3.496	3.667	3.895	4.218	4.718	5.614	7.823
25	3.129	3.217	3.324	3.457	3.627	3.855	4.177	4.675	5.568	7.770
26	3.094	3.182	3.288	3.421	3.591	3.818	4.140	4.637	5.526	7.721
27	3.062	3.149	3.256	3.388	3.558	3.785	4.106	4.601	5.488	7.677
28	3.032	3.120	3.226	3.358	3.528	3.754	4.074	4.568	5.453	7.636
29	3.005	3.092	3.198	3.330	3.499	3.725	4.045	4.538	5.420	7.598
30	2.979	3.067	3.173	3.305	3.473	3.699	4.018	4.510	5.390	7.562
31	2.955	3.043	3.149	3.281	3.449	3.675	3.993	4.484	5.362	7.530
32	2.934	3.021	3.127	3.258	3.427	3.652	3.969	4.459	5.336	7.499
33	2.913	3.000	3.106	3.238	3.406	3.630	3.948	4.437	5.312	7.471
34	2.894	2.981	3.087	3.218	3.386	3.611	3.927	4.416	5.289	7.444
35	2.876	2.963	3.069	3.200	3.368	3.592	3.908	4.396	5.268	7.419
36	2.859	2.946	3.052	3.183	3.351	3.574	3.890	4.377	5.248	7.396
37	2.843	2.930	3.036	3.167	3.334	3.558	3.873	4.360	5.229	7.373
38	2.828	2.915	3.021	3.152	3.319	3.542	3.858	4.343	5.211	7.353
39	2.814	2.901	3.006	3.137	3.305	3.528	3.843	4.327	5.194	7.333
40	2.801	2.888	2.993	3.124	3.291	3.514	3.828	4.313	5.179	7.314
41	2.788	2.875	2.980	3.111	3.278	3.501	3.815	4.299	5.163	7.296
42	2.776	2.863	2.968	3.099	3.266	3.488	3.802	4.285	5.149	7.280
43	2.764	2.851	2.957	3.087	3.254	3.476	3.790	4.273	5.136	7.264
44	2.754	2.840	2.946	3.076	3.243	3.465	3.778	4.261	5.123	7.248
45	2.743	2.830	2.935	3.066	3.232	3.454	3.767	4.249	5.110	7.234
46	2.733	2.820	2.925	3.056	3.222	3.444	3.757	4.238	5.099	7.220
47	2.724	2.811	2.916	3.046	3.213	3.434	3.747	4.228	5.087	7.207
48	2.715	2.802	2.907	3.037	3.204	3.425	3.737	4.218	5.077	7.194
49	2.706	2.793	2.898	3.028	3.195	3.416	3.728	4.208	5.066	7.182
50	2.698	2.785	2.890	3.020	3.186	3.408	3.720	4.199	5.057	7.171
51	2.690	2.777	2.882	3.012	3.178	3.400	3.711	4.191	5.047	7.159

52	7.149	5.038	4.182	3.703	3.392	3.171	3.005	2.874	2.769	2.683
53	7.139	5.030	4.174	3.695	3.384	3.163	2.997	2.867	2.762	2.675
54	7.129	5.021	4.167	3.688	3.377	3.156	2.990	2.860	2.755	2.668
55	7.119	5.013	4.159	3.681	3.370	3.149	2.983	2.853	2.748	2.662
56	7.110	5.006	4.152	3.674	3.363	3.143	2.977	2.847	2.742	2.655
57	7.102	4.998	4.145	3.667	3.357	3.136	2.971	2.841	2.736	2.649
58	7.093	4.991	4.138	3.661	3.351	3.130	2.965	2.835	2.730	2.643
59	7.085	4.984	4.132	3.655	3.345	3.124	2.959	2.829	2.724	2.637
60	7.077	4.977	4.126	3.649	3.339	3.119	2.953	2.823	2.718	2.632
61	7.070	4.971	4.120	3.643	3.333	3.113	2.948	2.818	2.713	2.626
62	7.062	4.965	4.114	3.638	3.328	3.108	2.942	2.813	2.708	2.621
63	7.055	4.959	4.109	3.632	3.323	3.103	2.937	2.808	2.703	2.616
64	7.048	4.953	4.103	3.627	3.318	3.098	2.932	2.803	2.698	2.611
65	7.042	4.947	4.098	3.622	3.313	3.093	2.928	2.798	2.693	2.607
66	7.035	4.942	4.093	3.618	3.308	3.088	2.923	2.793	2.689	2.602
67	7.029	4.937	4.088	3.613	3.304	3.084	2.919	2.789	2.684	2.598
68	7.023	4.932	4.083	3.608	3.299	3.080	2.914	2.785	2.680	2.593
69	7.017	4.927	4.079	3.604	3.295	3.075	2.910	2.781	2.676	2.589
70	7.011	4.922	4.074	3.600	3.291	3.071	2.906	2.777	2.672	2.585
71	7.006	4.917	4.070	3.596	3.287	3.067	2.902	2.773	2.668	2.581
72	7.001	4.913	4.066	3.591	3.283	3.063	2.898	2.769	2.664	2.578
73	6.995	4.908	4.062	3.588	3.279	3.060	2.895	2.765	2.660	2.574
74	6.990	4.904	4.058	3.584	3.275	3.056	2.891	2.762	2.657	2.570
75	6.985	4.900	4.054	3.580	3.272	3.052	2.887	2.758	2.653	2.567
76	6.981	4.896	4.050	3.577	3.268	3.049	2.884	2.755	2.650	2.563
77	6.976	4.892	4.047	3.573	3.265	3.046	2.881	2.751	2.647	2.560
78	6.971	4.888	4.043	3.570	3.261	3.042	2.877	2.748	2.644	2.557

(continued)

79	6.967	4.884	4.040	3.566	3.258	3.039	2.874	2.745	2.640	2.554
80	6.963	4.881	4.036	3.563	3.255	3.036	2.871	2.742	2.637	2.551
81	6.958	4.877	4.033	3.560	3.252	3.033	2.868	2.739	2.634	2.548
82	6.954	4.874	4.030	3.557	3.249	3.030	2.865	2.736	2.632	2.545
83	6.950	4.870	4.027	3.554	3.246	3.027	2.863	2.733	2.629	2.542
84	6.947	4.867	4.024	3.551	3.243	3.025	2.860	2.731	2.626	2.539
85	6.943	4.864	4.021	3.548	3.240	3.022	2.857	2.728	2.623	2.537
86	6.939	4.861	4.018	3.545	3.238	3.019	2.854	2.725	2.621	2.534
87	6.935	4.858	4.015	3.543	3.235	3.017	2.852	2.723	2.618	2.532
88	6.932	4.855	4.012	3.540	3.233	3.014	2.849	2.720	2.616	2.529
89	6.928	4.852	4.010	3.538	3.230	3.012	2.847	2.718	2.613	2.527
90	6.925	4.849	4.007	3.535	3.228	3.009	2.845	2.715	2.611	2.524
91	6.922	4.846	4.004	3.533	3.225	3.007	2.842	2.713	2.609	2.522
92	6.919	4.844	4.002	3.530	3.223	3.004	2.840	2.711	2.606	2.520
93	6.915	4.841	3.999	3.528	3.221	3.002	2.838	2.709	2.604	2.518
94	6.912	4.838	3.997	3.525	3.218	3.000	2.835	2.706	2.602	2.515
95	6.909	4.836	3.995	3.523	3.216	2.998	2.833	2.704	2.600	2.513
96	6.906	4.833	3.992	3.521	3.214	2.996	2.831	2.702	2.598	2.511
97	6.904	4.831	3.990	3.519	3.212	2.994	2.829	2.700	2.596	2.509
98	6.901	4.829	3.988	3.517	3.210	2.992	2.827	2.698	2.594	2.507
99	6.898	4.826	3.986	3.515	3.208	2.990	2.825	2.696	2.592	2.505
100	6.895	4.824	3.984	3.513	3.206	2.988	2.823	2.694	2.590	2.503

Printed with permission from NIST/SEMATECH e-Handbook of Statistical Methods, http://www.itl.nist.gov/div898/handbook/, 12 May 2019

Note: *df* for higher variance should be looked row wise

F-Test Table – 1% Point (11–20)

	11	12	13	14	15	16	17	18	19	20
1	6083.35	6106.35	6125.86	6142.70	6157.28	6170.12	6181.42	6191.52	6200.58	6208.74
2	99.408	99.416	99.422	99.428	99.432	99.437	99.440	99.444	99.447	99.449
3	27.133	27.052	26.983	26.924	26.872	26.827	26.787	26.751	26.719	26.690
4	14.452	14.374	14.307	14.249	14.198	14.154	14.115	14.080	14.048	14.020
5	9.963	9.888	9.825	9.770	9.722	9.680	9.643	9.610	9.580	9.553
6	7.790	7.718	7.657	7.605	7.559	7.519	7.483	7.451	7.422	7.396
7	6.538	6.469	6.410	6.359	6.314	6.275	6.240	6.209	6.181	6.155
8	5.734	5.667	5.609	5.559	5.515	5.477	5.442	5.412	5.384	5.359
9	5.178	5.111	5.055	5.005	4.962	4.924	4.890	4.860	4.833	4.808
10	4.772	4.706	4.650	4.601	4.558	4.520	4.487	4.457	4.430	4.405
11	4.462	4.397	4.342	4.293	4.251	4.213	4.180	4.150	4.123	4.099
12	4.220	4.155	4.100	4.052	4.010	3.972	3.939	3.909	3.883	3.858
13	4.025	3.960	3.905	3.857	3.815	3.778	3.745	3.716	3.689	3.665
14	3.864	3.800	3.745	3.698	3.656	3.619	3.586	3.556	3.529	3.505
15	3.730	3.666	3.612	3.564	3.522	3.485	3.452	3.423	3.396	3.372
16	3.616	3.553	3.498	3.451	3.409	3.372	3.339	3.310	3.283	3.259
17	3.519	3.455	3.401	3.353	3.312	3.275	3.242	3.212	3.186	3.162
18	3.434	3.371	3.316	3.269	3.227	3.190	3.158	3.128	3.101	3.077
19	3.360	3.297	3.242	3.195	3.153	3.116	3.084	3.054	3.027	3.003
20	3.294	3.231	3.177	3.130	3.088	3.051	3.018	2.989	2.962	2.938
21	3.236	3.173	3.119	3.072	3.030	2.993	2.960	2.931	2.904	2.880
22	3.184	3.121	3.067	3.019	2.978	2.941	2.908	2.879	2.852	2.827
23	3.137	3.074	3.020	2.973	2.931	2.894	2.861	2.832	2.805	2.781

(continued)

24	2.738	2.762	2.789	2.819	2.852	2.889	2.930	2.977	3.032	3.094
25	2.699	2.724	2.751	2.780	2.813	2.850	2.892	2.939	2.993	3.056
26	2.664	2.688	2.715	2.745	2.778	2.815	2.857	2.904	2.958	3.021
27	2.632	2.656	2.683	2.713	2.746	2.783	2.824	2.871	2.926	2.988
28	2.602	2.626	2.653	2.683	2.716	2.753	2.795	2.842	2.896	2.959
29	2.574	2.599	2.626	2.656	2.689	2.726	2.767	2.814	2.868	2.931
30	2.549	2.573	2.600	2.630	2.663	2.700	2.742	2.789	2.843	2.906
31	2.525	2.550	2.577	2.606	2.640	2.677	2.718	2.765	2.820	2.882
32	2.503	2.527	2.555	2.584	2.618	2.655	2.696	2.744	2.798	2.860
33	2.482	2.507	2.534	2.564	2.597	2.634	2.676	2.723	2.777	2.840
34	2.463	2.488	2.515	2.545	2.578	2.615	2.657	2.704	2.758	2.821
35	2.445	2.470	2.497	2.527	2.560	2.597	2.639	2.686	2.740	2.803
36	2.428	2.453	2.480	2.510	2.543	2.580	2.622	2.669	2.723	2.786
37	2.412	2.437	2.464	2.494	2.527	2.564	2.606	2.653	2.707	2.770
38	2.397	2.421	2.449	2.479	2.512	2.549	2.591	2.638	2.692	2.755
39	2.382	2.407	2.434	2.465	2.498	2.535	2.577	2.624	2.678	2.741
40	2.369	2.394	2.421	2.451	2.484	2.522	2.563	2.611	2.665	2.727
41	2.356	2.381	2.408	2.438	2.472	2.509	2.551	2.598	2.652	2.715
42	2.344	2.369	2.396	2.426	2.460	2.497	2.539	2.586	2.640	2.703
43	2.332	2.357	2.385	2.415	2.448	2.485	2.527	2.575	2.629	2.691
44	2.321	2.346	2.374	2.404	2.437	2.475	2.516	2.564	2.618	2.680
45	2.311	2.336	2.363	2.393	2.427	2.464	2.506	2.553	2.608	2.670
46	2.301	2.326	2.353	2.384	2.417	2.454	2.496	2.544	2.598	2.660
47	2.291	2.316	2.344	2.374	2.408	2.445	2.487	2.534	2.588	2.651
48	2.282	2.307	2.335	2.365	2.399	2.436	2.478	2.525	2.579	2.642
49	2.274	2.299	2.326	2.356	2.390	2.427	2.469	2.517	2.571	2.633
50	2.265	2.290	2.318	2.348	2.382	2.419	2.461	2.508	2.562	2.625
51	2.257	2.282	2.310	2.340	2.374	2.411	2.453	2.500	2.555	2.617

52	2.610	2.547	2.493	2.445	2.403	2.366	2.333	2.302	2.275	2.250
53	2.602	2.540	2.486	2.438	2.396	2.359	2.325	2.295	2.267	2.242
54	2.595	2.533	2.479	2.431	2.389	2.352	2.318	2.288	2.260	2.235
55	2.589	2.526	2.472	2.424	2.382	2.345	2.311	2.281	2.253	2.228
56	2.582	2.520	2.465	2.418	2.376	2.339	2.305	2.275	2.247	2.222
57	2.576	2.513	2.459	2.412	2.370	2.332	2.299	2.268	2.241	2.215
58	2.570	2.507	2.453	2.406	2.364	2.326	2.293	2.262	2.235	2.209
59	2.564	2.502	2.447	2.400	2.358	2.320	2.287	2.256	2.229	2.203
60	2.559	2.496	2.442	2.394	2.352	2.315	2.281	2.251	2.223	2.198
61	2.553	2.491	2.436	2.389	2.347	2.309	2.276	2.245	2.218	2.192
62	2.548	2.486	2.431	2.384	2.342	2.304	2.270	2.240	2.212	2.187
63	2.543	2.481	2.426	2.379	2.337	2.299	2.265	2.235	2.207	2.182
64	2.538	2.476	2.421	2.374	2.332	2.294	2.260	2.230	2.202	2.177
65	2.534	2.471	2.417	2.369	2.327	2.289	2.256	2.225	2.198	2.172
66	2.529	2.466	2.412	2.365	2.322	2.285	2.251	2.221	2.193	2.168
67	2.525	2.462	2.408	2.360	2.318	2.280	2.247	2.216	2.188	2.163
68	2.520	2.458	2.403	2.356	2.314	2.276	2.242	2.212	2.184	2.159
69	2.516	2.454	2.399	2.352	2.310	2.272	2.238	2.208	2.180	2.155
70	2.512	2.450	2.395	2.348	2.306	2.268	2.234	2.204	2.176	2.150
71	2.508	2.446	2.391	2.344	2.302	2.264	2.230	2.200	2.172	2.146
72	2.504	2.442	2.388	2.340	2.298	2.260	2.226	2.196	2.168	2.143
73	2.501	2.438	2.384	2.336	2.294	2.256	2.223	2.192	2.164	2.139
74	2.497	2.435	2.380	2.333	2.290	2.253	2.219	2.188	2.161	2.135
75	2.494	2.431	2.377	2.329	2.287	2.249	2.215	2.185	2.157	2.132
76	2.490	2.428	2.373	2.326	2.284	2.246	2.212	2.181	2.154	2.128
77	2.487	2.424	2.370	2.322	2.280	2.243	2.209	2.178	2.150	2.125
78	2.484	2.421	2.367	2.319	2.277	2.239	2.206	2.175	2.147	2.122

(continued)

79	2.481	2.418	2.364	2.316	2.274	2.236	2.202	2.172	2.144	2.118
80	2.478	2.415	2.361	2.313	2.271	2.233	2.199	2.169	2.141	2.115
81	2.475	2.412	2.358	2.310	2.268	2.230	2.196	2.166	2.138	2.112
82	2.472	2.409	2.355	2.307	2.265	2.227	2.193	2.163	2.135	2.109
83	2.469	2.406	2.352	2.304	2.262	2.224	2.191	2.160	2.132	2.106
84	2.466	2.404	2.349	2.302	2.259	2.222	2.188	2.157	2.129	2.104
85	2.464	2.401	2.347	2.299	2.257	2.219	2.185	2.154	2.126	2.101
86	2.461	2.398	2.344	2.296	2.254	2.216	2.182	2.152	2.124	2.098
87	2.459	2.396	2.342	2.294	2.252	2.214	2.180	2.149	2.121	2.096
88	2.456	2.393	2.339	2.291	2.249	2.211	2.177	2.147	2.119	2.093
89	2.454	2.391	2.337	2.289	2.247	2.209	2.175	2.144	2.116	2.091
90	2.451	2.389	2.334	2.286	2.244	2.206	2.172	2.142	2.114	2.088
91	2.449	2.386	2.332	2.284	2.242	2.204	2.170	2.139	2.111	2.086
92	2.447	2.384	2.330	2.282	2.240	2.202	2.168	2.137	2.109	2.083
93	2.444	2.382	2.327	2.280	2.237	2.200	2.166	2.135	2.107	2.081
94	2.442	2.380	2.325	2.277	2.235	2.197	2.163	2.133	2.105	2.079
95	2.440	2.378	2.323	2.275	2.233	2.195	2.161	2.130	2.102	2.077
96	2.438	2.375	2.321	2.273	2.231	2.193	2.159	2.128	2.100	2.075
97	2.436	2.373	2.319	2.271	2.229	2.191	2.157	2.126	2.098	2.073
98	2.434	2.371	2.317	2.269	2.227	2.189	2.155	2.124	2.096	2.071
99	2.432	2.369	2.315	2.267	2.225	2.187	2.153	2.122	2.094	2.069
100	2.430	2.368	2.313	2.265	2.223	2.185	2.151	2.120	2.092	2.067

Printed with permission from NIST/SEMATECH e-Handbook of Statistical Methods, http://www.itl.nist.gov/div898/handbook/, 12 May 2019
Note: df for higher variance should be looked row wise

Appendix 6: Data of File 'psoriasis.csv' Used in Chap. 29

S. no.	Age	Sex	Smoker	Alcoholic	Stress	Duration(m)	Location	PASI	WBC	RBC	HB	HCT	MCV	MCH	MCHC	PLT
1	13	M	0	0	2	4	W	6	7.02	4.75	13	39.5	83.2	27.4	32.9	272
2	21	M	0	0	3	12	W	4.3	10.02	5.62	15.7	46.3	82.4	27.9	33.9	277
3	30	F	0	0	2	120	W	11.3	8.61	3.89	12.3	38.3	98.5	31.6	32.1	352
4	18	M	0	0	2	8	W	20.3	8.43	4.56	14.6	43.7	95.8	32	33.4	259
5	33	M	1	1	3	24	W	7.2	10.26	5.11	16.3	46.7	91.4	31.9	34.9	238
6	25	M	1	1	2	4	W	6	6.82	4.59	13.7	40.5	88.2	29.8	33.8	180
7	18	F	0	0	4	1	W	34.3	7.94	3.85	12.3	38.4	99.7	31.9	32	307
8	35	M	1	0	4	24	W	44.6	4.21	3.51	10.7	34.8	99.1	30.5	30.7	386
9	65	M	1	1	2	24	W	61.8	2.23	4.45	12.8	40.8	91.7	28.8	31.4	140
10	60	M	0	0	3	120	W	53	3.76	4.32	14.2	44.3	102.5	32.9	32.1	302
11	13	F	0	0	4	12	W	1.8	6.86	4.22	10.7	34.8	82.5	25.4	30.7	219
12	23	M	1	0	2	24	W	2.8	20.64	5.94	14.2	45.1	75.9	23.9	31.5	216
13	26	F	0	0	2	36	W	10.4	11.46	4.23	12.7	40.5	95.7	30	31.4	211
14	40	M	1	1	2	120	W	3.9	10.22	5.59	16.9	49.8	89.1	30.2	33.9	209
15	28	F	0	0	2	2	W	0	5.67	4.13	12.3	38.9	94.2	29.8	31.6	278
16	33	M	0	0	4	6	W	25.8	6.73	5.05	14.9	42.1	83.4	29.5	35.4	373
17	70	M	0	0	3	120	W	13.3	6.43	3.34	11.3	35.9	107.5	33.8	31.5	299
18	45	F	0	0	2	36	W	0	8.5	4.3	11.9	39.1	90.9	27.7	30.4	295
19	45	M	1	0	3	120	W	16.9	10.17	4.86	15.7	46.6	95.9	32.3	33.7	241
20	57	F	0	0	2	4	W	2.5	4.53	3.66	12.4	39.5	107.9	33.9	31.4	224
21	50	F	0	0	2	6	W	0	6.93	3.77	11.5	36.9	97.9	30.5	31.2	226
22	13	M	0	0	2	3	W	0	4.88	4.65	14.5	42.8	92	31.2	33.9	214
23	21	F	0	0	2	18	W	12.2	8.31	4.35	11	34.8	80	25.3	31.6	365

(continued)

S. no.	Age	Sex	Smoker	Alcoholic	Stress	Duration(m)	Location	PASI	WBC	RBC	HB	HCT	MCV	MCH	MCHC	PLT
24	50	M	1	1	3	7	W	11.7	7.46	3.66	13.1	39.8	108.7	35.8	32.9	215
25	48	M	1	0	4	120	W	15.5	12.29	4.98	14.7	44.5	89.4	29.5	33	184
26	22	M	0	0	3	60	W	1.8	5.42	3.29	10.5	29.7	90.3	31.9	35.4	145
27	90	M	0	0	3	120	W	15	6.84	4.57	14	43	94.1	30.6	32.6	202
28	40	F	0	0	2	120	W	3.2	7.83	2.97	7.5	25.2	84.8	25.3	29.8	130
29	38	F	0	0	3	120	W	21.8	11.18	3.57	13	35.6	99.7	36.4	36.5	262
30	64	M	1	1	3	48	W	25.2	6.66	4.69	15.2	41.2	87.8	32.4	36.9	173
31	21	M	1	1	3	36	W	19	10.84	4.42	15	41.1	93	33.9	36.5	302
32	26	M	1	1	3	6	W	17.2	10.6	3.71	14.6	38.5	103.8	39.4	37.9	185
33	18	M	0	0	3	18	W	9.6	8.06	4.69	16.2	43.1	91.9	34.5	37.6	254
34	65	M	0	1	3	48	W	8.1	7.3	3.54	12.1	34.4	97.2	34.2	35.2	204
35	25	M	1	1	3	48	W	21.9	7.52	4.5	15	43.5	96.7	33.3	34.5	300
36	40	M	1	0	2	24	W	0.7	6.02	4.75	17.1	47.3	99.6	36	36.2	210
37	27	F	0	0	1	24	W	0	5.53	3.74	11.9	34.6	92.5	31.8	34.4	154
38	16	F	0	0	2	36	W	4.4	6.66	4.1	12.3	36.4	88.8	30	33.8	389
39	47	M	0	0	2	1	W	2	5.79	4.35	16.7	45.5	104.6	38.4	36.7	184
40	25	F	0	0	4	48	W	7.3	11.47	3.75	11.1	34.2	91.2	29.6	32.5	299
41	46	F	0	0	4	36	W	11.4	9.69	4.4	12.4	39.4	89.5	28.2	31.5	295
42	27	M	0	0	4	3	W	0.6	5.59	5.36	12.1	37.5	70	22.6	32.3	203

Index

© Springer Nature Singapore Pte Ltd. 2019
S. K. Yadav et al., *Biomedical Statistics*,
https://doi.org/10.1007/978-981-32-9294-9

Printed in the United States
By Bookmasters